Theory and Applications of Variable Structure Systems

ACADEMIC PRESS RAPID MANUSCRIPT REPRODUCTION

Proceedings of a Seminar on Theory and Applications of Variable Structure Systems with Emphasis on Modeling and Identification, Held in Sorrento, Italy, April 4-7, 1972.

Theory and Applications of Variable Structure Systems

edited by

R. R. Mohler
Department of Electrical and Computer Engineering, Oregon State University
Corvallis, Oregon

A. Ruberti
Istituto di Automatica
University of Rome
Rome, Italy

Academic Press
New York and London
1972

ACADEMIC PRESS, INC.
111 Fifth Avenue, New York, New York 10003

United Kingdom Edition published by
ACADEMIC PRESS, INC. (LONDON) LTD.
24/28 Oval Road, London NW1

LIBRARY OF CONGRESS CATALOG CARD NUMBER: 72-88377

PRINTED IN THE UNITED STATES OF AMERICA

CONTENTS

CONTRIBUTORS

A. V. Balakrishnan, System Science Department, School of Engineering and Applied Science, University of California, Los Angeles, California 90024

G. Basile, Istituto di Automatica, Viale Risorgimento, 2 Bologna, Italy

E. Biondi, Istituto di Elettrotecnica ed Elettronica del Politecnico di Milano, No P. le L da Vinci, 32-20133 Milan, Italy

R. W. Brockett, Division of Engineering and Applied Physics, Harvard University, Cambridge, Massachusetts 02138

C. Bruni, Istituto di Automatica, Via Eudossiana, Universita di Roma, 18-00184 Rome, Italy

P. d'Alessandro, Istituto di Automatica, Via Eudossiana, Universita di Roma, 18-00184 Rome, Italy

G. Di Pillo, Centro di Studio dei Sistemi di Controlo e Calcolo Automatici, del CNR, Via Eudossiana, 18-00184 Rome, Italy

F. Donati, Istituto Elettrotecnico Politecnico di Torino, Cso Duca degli Abruzzi, 24-10129 Torino, Italy

A. Isidori, Istituto di Automatica, Via Eudossiana, Universita di Roma, 18-00184 Rome, Italy

H. K. Knudsen, Department of Electrical Engineering and Computer Science, University of New Mexico, Albuquerque, New Mexico 87106

G. Koch, Istituto di Automatica, Via Eudossiana, Universita di Roma, 18-00184 Rome, Italy

G. Marchesini, Istituto di Elettrotecnica ed Elettronica, Via Gradenigo, Universita di Padova, 6/a-35100 Padova, Italy

G. Marro, Universita di Genoa, Genoa, Italy

R. R. Mohler, Department of Electrical and Computer Engineering, Oregon State University, Corvallis, Oregon 97331

B. C. Patten, Department of Zoology, and Institute of Ecology, University of Georgia, Athens, Georgia 30601

G. Picci, Istituto di Elettrotecnica ed Elettronica, Via Gradenigo, Universita di Padova, 6/a - 35100 Padova, Italy

A. Ruberti, Istituto di Automatica, Via Eudossiana, Universita di Roma, 18-00184 Rome, Italy

R. Schmid, Istituto di Elettrotecnic ed Elettronica del Potitecnico di Milano, No P. le L da Vinci, 32-20133 Milan, Italy

B. P. Zeigler, Logic of Computers Group, Department of Computer and Communication Sciences, University of Michigan, Ann Arbor, Michigan 48104

PREFACE

The seminar for which the proceedings are published here was conceived from common research interests on bilinear systems both at the Electrical and Computer Engineering Department of Oregon State University and at the Istituto di Automatica of Rome University.

Bilinear systems frequently arise quite naturally. More generally, bilinear systems are a reasonable compromise between the conflicting demands of accuracy and simplicity in models. They make it possible to overcome, on the one hand, the limitations imposed by linear models and, on the other hand, the complications associated with more highly nonlinear systems.

The significance of bilinear systems stems from their so-called variable structure or adaptive nature which was the theme chosen for the seminar. From this base, the first seminar on Theory and Applications of Variable Structure Systems was organized with some emphasis given to bridging the notorious gap between theory and applications. It is apparent that many of the problems facing modern society involve complex interacting systems of variable structure rather than mathematically convenient systems. Practice dictates a need for further development of theory. At the same time, certain new theories suggest new approaches to better understand these complex systems.

Alternating emphasis on theory and applications may well be expected in a period of rapid transformation such as we live in. Indeed, theory and practice are interwoven throughout the development of science and technology. The recent proliferation of theoretical developments in system science is an indication of the existence of new problems which need to be set in a rational framework in order that they be dominated and solved.

A typical aspect of the present phase of automatic control is the broadening of investigation and research from traditional engineering and physical systems, such as industrial processes, to include biological systems, socio-economic systems, ecological ones, etc.

It is sufficient to glance through the list of papers to come across the two characteristics just outlined: "the theoretical development of formal models" and "the extension of the field of applications."

As regards the papers presented to the Seminar, it is clear that they cannot and do not claim to offer a complete and exhaustive picture. As for other meetings of this kind, they have their origin in the converging interests of certain research groups and the personal relations that exist among them. The papers can be divided into three groups according to subject. The first subject concerns the description and identification of variable structure systems with particular regard to bilinear systems. The second concerns the modeling of physiological and ecological systems. This point is characterized by the range and the variety of the topics, which shows by contrast, however, the validity and unifying power of the systems approach. The Seminar was broadened by a third subject which delves into control, but which is still fundamental to system science. Its presence completes the picture. In particular, the following problems are examined: evaluation of the uncertainty of the models of the processes to be controlled, deeper structural analysis of processes for the synthesis of multiple regulators, and synthesis and approximate implementation of optimal control.

We wish to thank the National Science Foundation and the Consiglio Nazionale delle Ricerche for their support of this Seminar in the framework of the Cultural Agreement between the United States and Italy. Also, we express our sincere thanks to all colleagues and friends who collaborated to make this a successful venture.

Theory and Applications of Variable Structure Systems

MODELLING AND IDENTIFICATION THEORY: A FLIGHT CONTROL APPLICATION

A.V. Balakrishnan†

Abstract

In this paper we consider Identification problems for linear (and bilinear) dynamic systems in the presence of 'state' disturbance as well as measurement error. A general theory of identification (motivated by the application) is developed and is illustrated by application to a specific problem in Flight Control. This problem has all the canonical features of any identification problem — but what is more, verification of the theory on actual (flight) data (as opposed to computer simulated data) shows excellent agreement. A feature of the theory is that Ito integrals are <u>not</u> used.

---

†Research supported in part under Gr. No. 68-1408, AFOSR Applied Math. Division, USAF.

1. The problem of identifying system parameters from measurements ('input-output' data) is of growing interest in many areas of application. For a comprehensive recent survey see [4]. In this paper we shall consider such a problem for linear (and bilinear) dynamic systems in the presence of 'state' disturbance as well as measurement error. A general theory of identification (motivated by the application) is developed and is illustrated by application to a specific problem in Flight Control. This problem has all the canonical features of any identification problem — but what is more, verification of the theory on actual (flight) data (as opposed to computer simulated data) shows excellent agreement.

One feature of the theory presented here is that Ito integrals are not invoked — we take a "white noise" point of view as opposed to the "Wiener process" point of view. Lest there be alarm, the necessary justification and relation to Ito integrals is given in the Appendix. There are many advantages to our approach — one is that it makes the transition from sampled (finite dimensional) model to continuous time model quite transparent. Secondly it is more in accord with the practical situation where white noise is really band-limited noise of bandwidth large compared to the signal, and our integrals can be viewed as an approximation to the Ito integral — and an approximation is necessary anyhow in practice.

2. Mathematical Theory of Identification

2.1 A One-Dimensional Example

In order to illustrate the ideas involved and not lose them in purely technical details, we begin the mathematical theory with a concrete one-dimensional example. Suppose then the observation is modelled by

$$y(t) = s(t) + n(t) \tag{2.1}$$

where the measurement error is represented by white Gaussian noise $n(t)$ with spectral density normalized to unity. [As usual we are assuming that errors of scaling or bias, etc., have been removed already.] Much depends on how well we can model the system output $s(t)$. In our example, let us take:

$$\dot{s}(t) = k\, s(t) + \alpha n_s(t) + u(t) \tag{2.2}$$

where k is an unknown constant, and $n_s(t)$ is white Gaussian (independent of the observation noise n(.)), with spectral density, also unity. The initial condition s(0) will in general also be unknown; however, we shall consider only systems which are stable, and times of observation long enough that the response to the initial condition is small relative to the response to the (known) input u(.); we shall, in addition, impose "Identifiability" conditions. The problem then is to 'identify' to parameters k and α from y(s), $0 < s < T$ and we will need to know in particular the dependence on T.

We begin by assuming that our model is perfectly correct. That is to say, the observation corresponds to a value of k and α, and to be specific, denote them by k_o, α_o. They are unknown parameters in principle, but in most instances we do know that they are in a reasonably small specifiable neighborhood ahead of time. We shall return to this point later. With this point of departure, we can now invoke the standard apparatus of parameter estimation of classical statistics. We seek 'consistent', 'unbiassed' estimates; actually we can only manage 'asymptotically unbiassed' estimates. Moreover, we have to specify a computing algorithm which operating on the data y(.) will yield a close enough approximation to such estimates.

We choose the method of maximum likelihood for selecting the estimates. Here is where some analysis comes in. It is helpful to first consider the case where the data is sampled, so that the observation is just a vector Y of finite dimensions. Let m(t) denote the response to the input u(.) in the absence of any noise. Then we can write (2.1) in the form:

$$Y = S + m + N \tag{2.3}$$

Here we use S to indicate the (state) response to state noise in the absence of input; let R denote its covariance matrix. The covariance matrix of the noise vector N is of course the identity matrix I. Let θ denote the parameter set or vector (two-dimensional in our example). Then we have (using [,] to denote dot products) for the 'log-likelihood":

$$\text{Log } P(Y/\theta) = (-\tfrac{1}{2})\Big\{[(I+R)^{-1}(Y-m),\ Y-m]-[Y,Y] + \text{Log Det } |I+R|\Big\} \tag{2.4}$$

where we have exploited the Gaussian assumptions. A similar formula holds in the continuous case; we need only to consider Y as a vector in $L_2(0,T)$; R is the integral operator with kernel given by R(s,t) where R(s,t) is the covariance of the process

$$\int_0^t e^{k(t-\sigma)} n_s(\sigma)\,d\sigma \tag{2.5}$$

and

$$m(t) = \int_0^t e^{k(t-s)} u(s)\,ds \tag{2.6}$$

and we interpret, as usual, the dot products in (2.4) as inner-products in $L_2(0,T)$. See Appendix for the theory involved; it is unorthodox in that Ito integrals are not used. But the important point now is that we can write:

$$(I+R)^{-1} = (I - \mathscr{L})^* (I - \mathscr{L}) \tag{2.7}$$

where $\mathscr{L}$ is a Volterra operator, and * denotes adjoint. Moreover we have the equally important result that:

$$\text{Log Det } |I+R| = \text{Trace}(\mathscr{L} + \mathscr{L}^*) \tag{2.8}$$

This is an application of the Factorization theorem of Krein; it has nothing to do with whether the process s(t) has a differential (or dynamic system representation as in (2.2)). On the other hand because (2.2) holds, we can observe that the operator $\mathscr{L}$ is given by the Bucy-Kalman filter. In fact in our example $\mathscr{L}$ is defined by:

$$\mathscr{L}f \sim \int_0^t \phi(t)\,\phi(s)^{-1}\,P(s)\,f(s)\,ds \tag{2.9}$$

where

$$\dot{\phi}(t) = (k-P(t))\ \phi(t)$$

and

$$\dot{P}(t) = kP(t) + P(t)\ k - P(t)^2 + \alpha^2;\ P(0) = 0 \quad (2.10)$$

Note that

$$(I - \mathscr{L})(Y - m)$$

is white noise. Finally, we can now rewrite (2.4) as:

$$(-\tfrac{1}{2}) \Big\{ [(I-\mathscr{L})(Y-m),\ (I-\mathscr{L})(Y-m)] - [Y,Y] + + 2\int_0^T P(t)\ dt \Big\}$$

since

$$\text{Tr.}\ (\mathscr{L}+\mathscr{L}^*) = 2\int_0^T P(t)dt$$

using (3.9). After a little arithmetic we can write (3.4) as:

$$(-\tfrac{1}{2}) \Big\{ ||m + \mathscr{L}(Y-m)||^2 - 2[Y,m + \mathscr{L}(Y-m)] + 2\int_0^T P(t)dt \Big\} \quad (2.11)$$

Finally, taking time averages, we can rewrite this as (omitting the (-1/2) factor at the same time):

$$q(\theta;T) = \frac{1}{T} \Big\{ ||m + \mathscr{L}(Y-m)||^2 - 2[Y,m + \mathscr{L}(Y-m)] \Big\} + \frac{2}{T}\int_0^T P(t)dt \quad (2.12)$$

For large T, observe that the third term can be approximated as:

$$\frac{1}{T}\int_0^T P(t)dt \approx P \quad (2.13)$$

where

$$P = \lim_{t \to \infty} P(t)$$

and can be obtained by setting the time derivative in (3.10) to be zero:

$$0 = 2\,kP - P^2 + \alpha^2$$

and hence is given by:

$$P = k + \sqrt{k^2 + \alpha^2}$$

Although theoretically, the maximum likelihood estimate is the one that maximizes (2.12), we note that a computer logarithm can only locate a local 'turning point'. Thus we now assume that we know an interval in which the unknown parameters lie. We look for a root of the gradient of (2.12) in such an interval; and in order to ensure a root we need some conditions on the second gradient — this yields our 'identifiability' conditions. In our example, the gradient of (2.12) is the vector:

$$\nabla_\theta q(\theta;T) = \begin{pmatrix} \dfrac{\partial}{\partial k} q(\theta;T) \\ \\ \dfrac{\partial}{\partial \alpha} q(\theta;T) \end{pmatrix} = \begin{pmatrix} (-2/T)\left[(I-\mathscr{L})(Y-m),\ (I-\mathscr{L})\dfrac{\partial m}{\partial k} + \dfrac{\partial \mathscr{L}}{\partial k}(Y-m)\right] + (2/T)\displaystyle\int_0^T \dfrac{\partial P(t)}{\partial k}\,dt \\ \\ (-2/T)\left[(I-\mathscr{L})(Y-m),\ \dfrac{\partial \mathscr{L}}{\partial \alpha}(Y-m)\right] + (2/T)\displaystyle\int_0^T \dfrac{\partial}{\partial \alpha} P(t)\,dt \end{pmatrix} \tag{2.14}$$

we note that because of the 'quadratic' form of (2.12), the second gradient matrix splits into two terms one of which is actually non-negative definite, and, the second goes to zero again at the true parameter values for large T. The non-negative definite part is:

$$Q(\theta;T) = \begin{matrix} \frac{2}{T}\left[(I-\mathscr{L})\frac{\partial m}{\partial k} + \frac{\partial \mathscr{L}}{\partial k}(Y-m),\ (I-\mathscr{L})\frac{\partial m}{\partial k} + \frac{\partial \mathscr{L}}{\partial k}(Y-m)\right] & \frac{2}{T}\left[\frac{\partial \mathscr{L}}{\partial \alpha}(Y-m),\ \frac{\partial \mathscr{L}}{\partial \alpha}(Y-m)\right] \\ \frac{2}{T}\left[\frac{\partial \mathscr{L}}{\partial \alpha}(Y-m),\ (I-\mathscr{L})\frac{\partial m}{\partial k} + \frac{\partial \mathscr{L}}{\partial k}(Y-m)\right] & \frac{2}{T}\left[(I-\mathscr{L})\frac{\partial m}{\partial k} + \frac{\partial \mathscr{L}}{\partial k}(Y-m),\ \frac{\partial \mathscr{L}}{\partial \alpha}(Y-m)\right] \end{matrix} \tag{2.15}$$

Our 'identifiability' condition is then that this matrix be positive definite as T goes to infinity at the true parameter values. Since in practice we do not know exact parameter values, we demand this be true in the finite interval in which they are known to lie. In particular in our example, denoting the limiting matrix by $Q(\theta_o)$, it can be shown with a little analysis that $Q(\theta_o)$ as singular if and only if

$$\frac{\partial P}{\partial \alpha} = 0 \tag{2.16}$$

This can happen only if $\alpha_o = 0$; or, the unknown α lies in an interval containing the origin, and this is an unnatural condition. Note that no condition on the input (other than the existence of time-averages) is used; in particular the input can be identically zero. So long as α lies in an interval not including the origin, the identifiability conditions are satisfied. But if there is a nonzero input, it 'helps' in the sense that $Q(\theta_o)$ is "more" positive and hence the functional we are minimizing 'less flat' in the interval of interest. Note that $Q(\theta_o)$ is nonsingular if and only if the simpler matrix Q is: (see (2.21))

$$Q = \begin{bmatrix} \left[\frac{\partial \tilde{\mathscr{L}}}{\partial k}, \frac{\partial \tilde{\mathscr{L}}}{\partial k}\right]_\infty & \left[\frac{\partial \tilde{\mathscr{L}}}{\partial \alpha}, \frac{\partial \tilde{\mathscr{L}}}{\partial k}\right]_\infty \\ \left[\frac{\partial \tilde{\mathscr{L}}}{\partial \alpha}, \frac{\partial \tilde{\mathscr{L}}}{\partial k}\right]_\infty & \left[\frac{\partial \tilde{\mathscr{L}}}{\partial \alpha}, \frac{\partial \tilde{\mathscr{L}}}{\alpha \alpha}\right]_\infty \end{bmatrix} \tag{2.17}$$

A computer algorithm that can be used is:

$$\theta_{n+1} = \theta_n - Q(\theta_n;T)^{-1} \nabla_\theta \, q(\theta_n;T) \tag{2.18}$$

This algorithm can be shown to converge for initial choice in a small enough neighborhood for all T sufficiently large. The main thing to note about the algorithm is that it does not invoke second-derivatives in the matrix. Even further simplification is possible for large enough T by dropping additional terms in the matrix.

Continuing the one dimensional example let us look at a nonlinear case, that is where the state-dynamics are nonlinear. Of course if there is no state noise, the problem is trivial. On the other hand with state noise, the general problem would appear to be hopeless at the present time in the sense no comparable theory or computational approximation is known at the present time. Fortunately there is one case where we can apply the previous considerations — and this case is of great importance apparently, judging by the interest in this Conference. This is the bilinear case, which in one dimension we shall write, generalizing (2.2):

$$\dot{s}(t) = k\, s(t) + b\, u(t)\, s(t) + c\, u(t) + \alpha n_s(t)$$

Now, the main point is that since the control u(t) is known, we have a linear, albeit time-varying system which we shall rewrite as:

$$s(t) = a(t)\, s(t) + c\, u(t) + \alpha n_s(t); \quad a(t) = k+b\, u(t)$$

Then (2.3) can still be used, with the interpretation that

$$m(t) = a(t)\, m(t) + c\, u(t)$$

and

$$S \sim s(t) - m(t)$$

so that (2.11) still applies with

$$\mathscr{L} f \sim \phi(t) \int_0^t \phi(s)^{-1} P(s)\ f(s) ds$$

$$\dot{\phi}(t) = (a(t) - P(t))\ \phi(t)$$

$$\dot{P}(t) = 2a(t)\ P(t) - P(t)^2 + \alpha^2 ; P(0) = 0$$

We can then proceed to find a root of the gradient of (2.12) for fixed T, with respect to the parameters. However, we cannot employ (2.13), and in fact cannot say much about what happens as $T \to \infty$, and no statements about 'consistency' or 'asymptotic unbiassedness' unless the input u(t) is restricted further.

2.2 General results for linear dynamic systems with state noise: Referring all the details and proofs to [3], we can now state the main results for the general case of a linear dynamic system with unknown parameters. As before, θ will denote the vector of unknown parameters. We deliberately avoid the use of Ito integrals. For the system

$$\begin{aligned} \dot{x}(t) &= A\ x(t) + B\ u(t) + F\ n(t) \\ y(t) &= C\ x(t) + D\ u(t) + G\ n(t) \end{aligned} \tag{2.19}$$

where n(t) is white Gaussian unit spectral density,

$$FG^* = 0$$

(state noise independent of observation noise), u(t) is a known input with some conditions on it as stipulated below, and the unknown parameters are present in any or all of the matrices A,B,C,D,F, but not in G. Let $m(\theta;t)$ denote the response in the absence of all noise (i.e., when n(t) is zero). We assume that GG* is the identity (since this only involves an appropriate matrix multiplication otherwise). The operator $\mathscr{L}(\theta)$ is now given by:

$$\mathscr{L}(\theta)\ f \sim C\ \phi(t) \int_0^t \phi(s)^{-1}\ P(s) C^*\ f(s) ds$$

where

$$\dot{\phi}(t) = (A-P(t)C^*C)\ \phi(t)$$

$$\dot{P}(t) = A\ P(t) + P(t)A^* + FF^* - P(t)C^*C\ P(t);\ P(0) = 0$$

Then the computing algorithm becomes:

$$\theta_{n+1} = \theta_n - Q(\theta_n;T)^{-1}\ \nabla_\theta\ q(\theta_n;T)$$

where $\nabla_\theta\ q(\theta;T)$ is the column vector

$$(-1)(2/T)\left\{[(I-\mathscr{L}(\theta_n))(Y-m),\ (I-\mathscr{L})\nabla_\theta m + (\nabla_\theta \mathscr{L})(Y-m)]\right.$$
$$\left.\int_0^T \nabla_\theta\ \mathrm{Tr.}\ CP(s)C^*ds\right\}$$

The matrix $Q(\theta;T)$ has the components:

$$(2/T)\ [(I-\mathscr{L}(\theta))\ m_i + \mathscr{L}_i(\theta)(Y-m),$$
$$(I-\mathscr{L}(\theta))\ m_j + \mathscr{L}_j(\theta)(Y-m)]$$

where the subscripts indicate partial derivatives with respect to components of θ. We assume that the input u(.) satisfies the condition that

$$\lim_{T\to\infty}\ (1/T)\int_0^T u(t)\ u(t+s)^*\ dt$$

exists, is finite, and defines a continuous function of s. We also assume that A is stable (all eigenvalues have negative real parts) and C-A is observable. Under these conditions P(t) converges as $t \to \infty$; call the limit matrix P. The processes thus become asymptotically stationary. Let $\tilde{\mathscr{L}}$ denote the stationary version of $\mathscr{L}$; that is

$$\tilde{\mathscr{L}}f = g;\ g(t) = \int_0^t e^{(A-PC^*C)(t-s)}PC^*\ f(s)ds$$

Then (at the true value)

$$\lim_{T\to\infty}\ Q(\theta;T)$$

is the matrix with components:

$$2[(I-\tilde{\mathscr{L}}(\theta))\, m_i,\ (I-\tilde{\mathscr{L}}(\theta))\, m_j]_{av.}$$

$$+\ 2[\tilde{\mathscr{L}}_i(\theta)(I+K(\theta)),\ \tilde{\mathscr{L}}_j(\theta)(I+K(\theta))]_\infty$$

where

$$I + K(\theta) = (I-\tilde{\mathscr{L}}(\theta))^{-1}$$

and

$$[f,g]_{av} = \lim \frac{1}{T}\int_0^T [f(t),\ g(t)]dt$$

$$[L_1,L_2]_\infty = \int_0^\infty [W_1(t),\ W_2(t)]dt \tag{2.20}$$

where

$$L_1 f = g; \quad g(t) = \int_0^t W_1(t-s)f(s)ds$$

and similarly for L_2, and the (letter) subscripts denote partial derivatives with respect to the components of θ. The identifiability condition is that this matrix be positive definite; or equivalently, if the matrix

$$[m_i,m_j]_{av.} + [\tilde{\mathscr{L}}_i(\theta),\ \tilde{\mathscr{L}}_j(\theta)]_\infty \tag{2.21}$$

be positive definite. See [3].

2.3 General Bilinear System with State Noise

For the sake of completeness we shall indicate the extension to the general bilinear case with state noise. Thus, we take the state equations:

$$\dot{x}(t) = A\ x(t) + B_1\ u(t))\ x(t) + B_2\ u(t) + F\ u(t)$$

and observation:

$$y(t) = C\ x(t) + D\ u(t) + G\ n(t)$$

where n(t) is white Gaussian with unit spectral density and

$$F\ G^* = 0$$

and where u(t) is a known input and unknown parameters are present in any of the coefficient matrices except G and for simplicity of notation we assume GG* is the identity.

$$A(t) = A + B_1\ u(t)$$

and let x(0) denote the unknown initial state. Let

$$\dot{m}_1(t) = A(t)\ m_1(t) + B_2\ u(t); \quad m_1(0) = x(0)$$

Let

$$m(t) = m_1(t) + D\ u(t)$$

Since the system is time-varying, we cannot afford to omit the response due to the initial conditions and we simply include that among the parameters to be determined. Continuing as before, let $\mathscr{L}(\theta)$ denote the operator:

$$\mathscr{L}\theta f \sim C\ \phi(t) \int_0^t \phi(s)^{-1}\ P(s)\ C^*\ f(s)ds$$

where

$$\dot{\phi}(t) = A(t)\ \phi(t)$$

$$\dot{P}(t) = A(t)\ P(t) + P(t)\ A(t)^*$$

$$+ FF^* - P(t)\ C^*C\ P(t); \quad P(0) = 0$$

Then the 'maximal likelihood' estimate is a root of the gradient [with respect to the unknown parameter vector θ, which now includes also the initial state x(0)] of:

$$q(\theta;T) = \left(-\frac{1}{2}\right)\left\{||m+\mathscr{L}(Y-m)||^2 - 2[Y,m+\mathscr{L}(Y-m)] + 2\int_0^T \mathrm{Tr.}C\ P(t)C^*dt\right\}$$

The expression for the gradient is the same as before, and the computing algorithm can also be taken the same as before. If the input u(t) is such that

$$\lim_{t \to \infty} A(t) = A + B\ u(\infty)$$

and the limit matrix is stable, then we can invoke the identifiability conditions as before, and under these conditions we have the asymptotic consistency of the estimates.

3. Applications: Flight Control Example

The Specific Flight Control Problem treated is that of designing stability augmentation systems for aircraft in wind-gust turbulence [1]. A nominal flight path is given and the air-craft parameters describing perturbation dynamics about the nominal change in a nonpredictable fashion. The designer is to determine a control logic which operates on available information such that the design specifications are met throughout the required mission.

The first discernible step is the 'modelling' — 'mathematical' [or 'calculable'] relationships between the control variables, the measurable variables and the desired performance criteria. It is of interest to note that even in a perfectly well-understood (in comparison to a biological system, let us say) physical system as aircraft flight dynamics this is quite complex. Being a physical system, the required relationships can be based on physical law — Newtonian mechanics in this case, and hence we can write down dynamic equations. [We remark paranthetically that in spite of all the "abstract system theory", 'dynamic' equations provide still the only means of such description. The use of automata for this purpose is still confined to textbooks.] Here again our experience must guide us to determine the level of complexity. For instance we can employ 'stability derivatives' — linearise the equations about the nominal, with appropriate partial derivatives describing the coefficients in the equation. But the aircraft has a flexible structure and the question of modelling structural response to wind-gust as well as the modelling of wind-gust itself leads to many imponderables and many compromises. It is easy enough to appreciate in a general way that although the 'real' system can be quite complicated, it may be that for the purposes of deducing the necessary control a much 'reduced' representation may be adequate. There is unfortunately no mathematical theory to guide us in this at the present time. So we take a lumped parameter model, and model the wind-gust as a stationary Gaussian process with a rational (but unknown) spectrum. See [1] for one version. We have then the 2nd phase of parameter identification. Here with the usual assumption of additive Gaussian noise we can make some progress using estimation theory and in particular the theory of stochastic differential systems. The control problem is by far the simpler one since once the system and noise dynamics are determined the 'regulator' feedback control theory [3] can be applied — at least assuming the control to be linear. The final phase is Adaptive control involving in particular identification in closed loop. Little is known at the present time about such systems in practice — even in theory many questions of optimization need to be resolved. Here we shall study only the identification problem which is admittedly the most important part. The linearized equations (longitudinal mode only) can be expressed in state-space formulation as:

$$\dot{x} = A\,x + B\,u + F\,n$$

$$y = C\,x + D\,u + E\,v + G\,n$$

Here the matrices A,B,C,D,F have the form:

$$A = \begin{bmatrix} Z_1 & 0 & 1 & Z_1 \\ 0 & 0 & 1 & 0 \\ M_1 & 0 & M_3 & M_1 \\ 0 & 0 & 0 & -\dfrac{\bar{v}}{1000} \end{bmatrix}$$

$$B = \begin{bmatrix} Z_4 \\ 0 \\ M_4 \\ 0 \end{bmatrix} \qquad F = \sigma \begin{bmatrix} 0 \\ 0 \\ 0 \\ \dfrac{1}{20\bar{v}} \end{bmatrix}$$

$$C = \begin{bmatrix} 0 & 0 & 1 & 0 \\ 0 & 1 & 0 & 0 \\ \dfrac{10M_1 - \bar{v}Z_1}{g} & 0 & \dfrac{10M_3}{g} & \dfrac{10M_1 - \bar{v}Z_1}{g} \\ k_1 & 0 & \dfrac{32k_1}{\bar{v}} & k_1 \end{bmatrix}$$

$$D = \begin{bmatrix} 0 \\ 0 \\ \dfrac{10M_4 - \bar{v}Z_4}{g} \\ 0 \end{bmatrix}$$

g = acceleration due to gravity

$E = 4\, \bar{v}\, Z_4$

$v(\cdot)$ is known (structural response)

$G = \text{diag.}\ [.0005, .0001, .01, .00001]$

The lettered entries are unknown, except for $\bar{v}$. The input u(.) is taken as a square wave, with an amplitude of 0.02 at a frequency of 0.4 cps. An example is presented in Fig. 1 & 2 with data given in Table I.

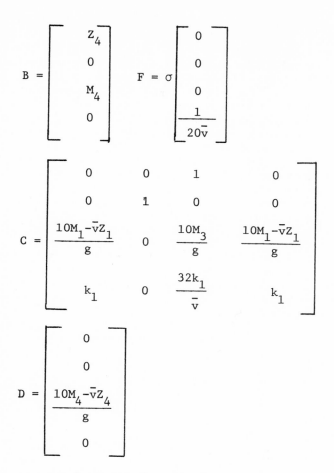

TABLE I

SIMULATION RESULTS

(All Calculations Performed by K. Iliff)

$$P = 10^{-8} \begin{vmatrix} 4.515 & 4.015 & 9.676 & -4.689 \\ 4.015 & 3.596 & 7.755 & -4.045 \\ 9.676 & 7.755 & 49.760 & -14.040 \\ -4.689 & -4.045 & -14.040 & 13.320 \end{vmatrix}$$

Parameter	True Value	Starting Value	Value Obtained After		
			1 Iteration	2 Iterations	3 Iterations
Z_1	-1.65	-2.4	-1.709	-1.666	-1.666
M_1	-54	-39	-58.90	-54.07	-54.07
M_3	-1.65	-2.4	-3.28	-1.288	-1.612
Z_4	-0.45	-0.675	-0.5818	-0.4607	-0.4549
M_4	-52.5	-36	-57.44	-52.47	-52.06
$(10^5)\sigma^2$	1.267	0.1	0.57	0.79	0.86

Length of Data 5 Seconds

Sampled at Intervals of .01 Sec.

$\bar{v}$ = 1670 f.p.s.

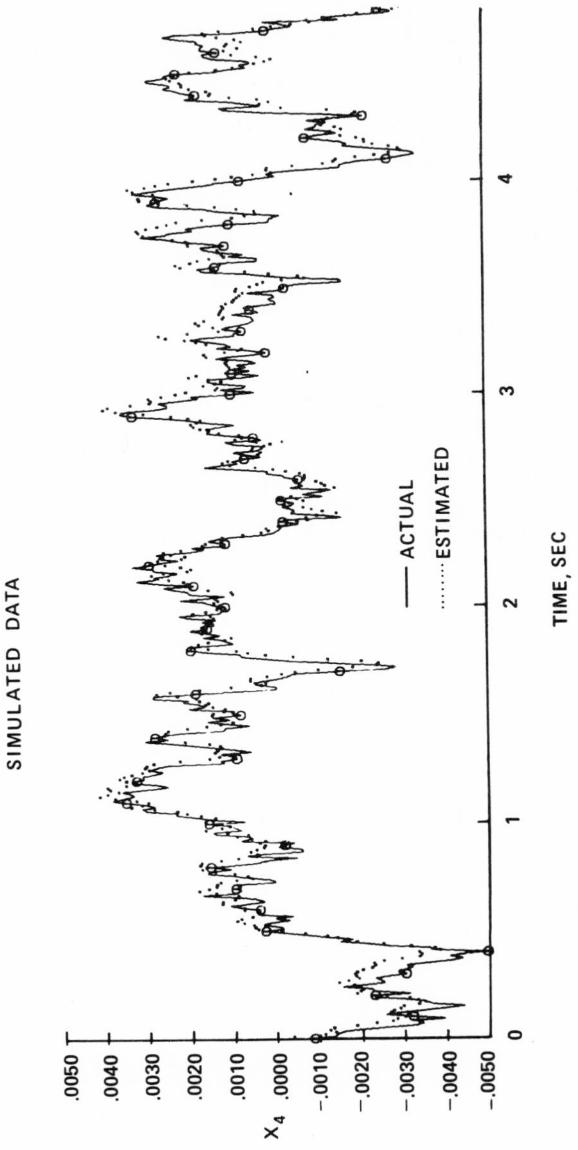

Fig. 1. Simulated comparison for Jet Star aircraft

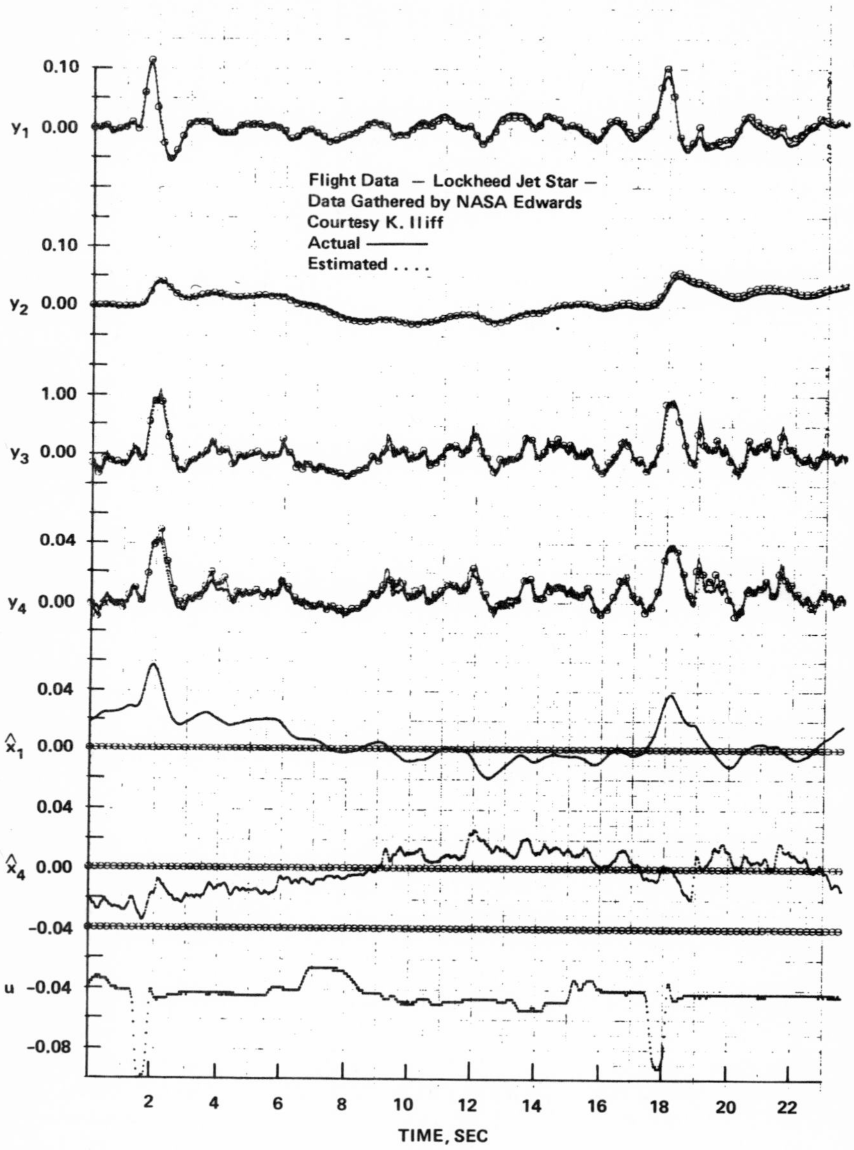

Fig. 2. Flight-test comparison of estimated and actual parameters.

APPENDIX

Our main purpose in this Appendix is to justify the use of 'white noise' as opposed to 'Wiener process'. The theory is complicated only in that we have to introduce probability measures which are only finitely additive but in return do not require the notion of the Ito integral. But beyond this technical detail there is a difference in point of view — on how to model 'white noise'. We claim that in terms of 'observed data' where the 'additive white noise' is really an approximation for the large bandwidth 'instrument error' term this view is perhaps closer to reality. It can be alternately explained as an approximation of the Ito integral — and an approximation necessary anyway to 'instrument' the Ito integral involving observed data.

White Noise

Let us first explain the model we shall use for white noise. Let [0,T] denote the time-interval of observation. Let $\underline{H} = L_2[0,T]$ denote the real Hilbert space over [0,T] of functions with range in E, Euclidean n-space. Let f(.) denote an element in $\underline{H}$. Then f(t) will be taken as n x 1 matrix function for each t, in the algebraic manipulations below. We introduce the 'Gauss' measure on $\underline{H}$. Let U denote a Borel set in E and let, for fixed g:

$$C = [f(.) \varepsilon H \mid [f,g] \varepsilon U, \ [g,g] = 1]$$

For each such 'cylinder' set we define:

$$P_G(C) = \int_U \frac{\exp - ||x||^2/2}{(2\pi)^{n/2}}$$

This yields a finitely additive measure on the cylinder sets of $\underline{H}$. This <u>cannot</u> be extended to be countably additive on $\underline{H}$. However we calculate the measure of any cylinder set. Very loosely stated, any ' observation' must be finite-dimensional, so this is not a real restriction. Let ω denote 'points' in $\underline{H}$. By ' white noise' we mean the "stochastic process":

$$n(t;\omega) = \omega(t)$$

Let $h(.)\varepsilon H$; consider the functional:

$$[h,n(.,\omega)]$$

This has its range in a finite dimensional space and inverse images of Borel sets are in $\mathscr{C}$. Moreover

$$E\left[e^{i[h,n(.)\omega)]}\right] = \int_{\underline{H}} e^{i[h,n(.)\omega)]} dpG = \exp^{-\frac{[h,h]}{2}}$$

Note that $[h,n(.;\omega)]$ is a Gaussian random variable. Let $\{\phi_k\}$ be a C.O.S. in $\underline{H}$. Let

$$\Phi_k(\omega) = [\phi_k, n(.;\omega)$$

Then for each ω, we have

$$n(.;\omega) = \sum_1^\infty \Phi_k(\omega)\phi_k$$

$$E[\Phi_k(\omega)^2] = 1$$

$$[n(.)\omega), n(.)\omega] = \sum_1^\infty \Phi_k(\omega)^2 \geq \sum_1^N \Phi_k(\omega)^2$$

Let P_m denote the projection operator corresponding to the space spanned by $\phi_1, .. \phi_m$. Then

$$P_m\; n(.;\omega)$$

has a finite dimensional range and is Borel measurable (inverse images of Borel sets are Borel sets in $\mathscr{C}$). Moreover

$$E[||P_m(n(.,\omega)||^2] = \sum_1^m E[\Phi_k(\omega)^2] = m$$

Hence

$$\lim_{m \to \infty} E[||P_m n(.,\omega)||^2] = \infty$$

Hence, even though the variable

$$||n(.,\omega)||^2$$

is <u>not</u> Borel measurable, the expected value would be infinite, even if it were.

Consider next the differential equation:

$$\dot{S}(t) = AS(t) + Fn(t;\omega); \; S(0) = 0$$

For each ω, this has a unique solution

$$S(t;\omega) = \int_0^t e^{A(t-\sigma)} Fn(\sigma;\omega)d\sigma$$

and moreover

$$S(.;\omega) = L\omega$$

where L is a linear bounded transformation of $\underline{H}$ into $\underline{H}$. Now

$$[h,S(.;\omega)] = [h,L\omega] = [L^*h,\omega]$$

is Gaussian for each h in $\underline{H}$. Moreover

$$\Phi(h) = E\left[e^{i[h,S(.)\omega)]}\right] = \exp - \frac{1}{2} [L^*h,L^*h]$$

We shall define a function $\Phi(.)$ mapping $\underline{H}$ into $\underline{H}$ 'measurable' if it is continuous in the topology induced by the seminorms: $||Sh||$, where S is any nonnegative definite trace-class operator.

Note the LL* is trace class. Hence $\Phi(h)$ is the characteristic functional of a countably additive probability measure on the Borel sets of H. In particular

$$||S(.,\omega)||^2$$

is measurable and

$$E[||S(.;\omega)||^2] = \text{Tr. } LL^*$$

Let L be any linear bounded transformating mapping $\mathscr{H}$ into a Hilbert Space. Then $L\omega$ is measurable if L^*L is trace-class. So is $[L\omega,\omega]$ if $L+L^*$ is trace-class. Next let us get back to (2.19). Note that m(t) is a function in $L_2[0,T]$ and that

$$y(t) = m(t) + C\int_0^t e^{A(t-\sigma)}F\, n(\sigma)d\sigma + G\, n(t)$$

yields a mapping of $L_2[0,T)$ into itself, which we can write as

$$y(\omega) = m(\cdot) + M\, n(.;\omega)$$

with $n(t;\omega)$ denoting white noise,

$$M f = g;\ g(t) = G\, f(t) + C\int_0^t e^{A(t-\sigma)}F\, n(\sigma;\omega)d\sigma$$

and hence, in the notation of Section 2 we see that

$$MM^* = (I + R)$$

Now use the Krein theorem to obtain

$$(I + R)^{-1} = (I - \mathscr{L})^* (I - \mathscr{L})$$

Then

$$E\left[e^{i[h,(I-\mathscr{L})(Y-m)]}\right] = \exp - [h,h]/2$$

so that

$$(I - \mathscr{L})\ (Y - m) \text{ is white noise.}$$

Moreover:

$$\int_H e^{i[h,y(\omega)]} dp_G$$

is the characteristic function of a finitely additive measure on the cylinder sets of $\underline{H}$. Call the corresponding measure p_y. Then p_y is 'absolutely continuous' with respect to p_G. Moreover the 'derivative' is given by the functional

$$h(Y) = \exp - \frac{1}{2}\left\{[(I-\mathscr{L})(Y-m),\ (I-\mathscr{L})(Y-m) - [Y,Y] + 2\int_0^T \text{Tr. } CP(t)C^*dt\right\};\ Y\varepsilon\ \underline{H}$$

Note that this functional, as we have seen, can be rewritten:

$$\text{Exp} - \frac{1}{2}\left\{||m+\mathscr{L}(Y-m)||^2 - 2[Y,m+\mathscr{L}(Y-m)] + 2\int_0^T \text{Tr. } CP(t)C^*dt\right\}$$

and measurability can be proven using this form because $(\mathscr{L}+\mathscr{L}^*)$ is a trace-class operator. We omit the details of this proof; as we have noted, it is a direct generalization of the finite-dimensional result. The relation to Ito integrals can be noted at this stage. Thus if we used Ito-integrals, the third term will not appear and

$$[Y,\ \mathscr{L}(Y-m)]$$

would become instead

$$[\mathscr{L}(dy-m),\ dy]$$

and the main point is that for Wiener process W(.) the Ito-integral

$$[\mathscr{L}dW,\ dW] \approx [\mathscr{L}n,n] - \frac{1}{2}\text{Trace } (\mathscr{L}+\mathscr{L}^*)$$

see [3, Chapter V]. Note in particular that

$$E[Y, \mathscr{L}(Y-m)] = \text{Tr. } (I+R) \frac{(\mathscr{L}+\mathscr{L}^*)}{2}$$

This continues to be true for any L such that L+L* is trace-class. Next, if L_1, L_2 are Hilbert-Schmidt operators,

$$EC[L_1(Y-m),\ L_2(Y-m)],\ = \text{Tr. } L_1(I+R)L_2^*$$

$$= \text{Tr. } (L_1(I-\mathscr{L})^{-1})(L_2(I-\mathscr{L})^{-1})^*$$

$$= \int_0^T dt \int_0^t [h_1(t;s),\ h_2(t,s)]ds$$

where $h_1(t;s)$ is defined by:

$$L_1(I-\mathscr{L})^{-1} f = g; \quad g(t) = \int_0^t h_1(t;s)ds$$

REFERENCES

1. L. Taylor and H. Rediess: "Flight Control Design Challenge", Proceedings, Joint Automatic Control Conference (JACC)-1970, Atlanta, Georgia.

2. A.V. Balakrishnan: "Identification and Adaptive Control: An Application to Flight Control Systems", Journal of Optimization Theory and Applications, June 1972.

3. A.V. Balakrishnan: "Stochastic Differential Systems: A Function Space Approach", Academic Press, Inc., 1972 (to be published).

4. K.J. Astrom and P. Eykhoff: Survey Paper: "System Identification", Proceedings of the IFAC Symposium on "Identification and Parameter Estimation", Prague, Czechoslovakia, 1970.

STRUCTURE ANALYSIS OF LINEAR AND BILINEAR DYNAMICAL SYSTEMS

P.d'Alessandro* - A.Isidori** - A.Ruberti***

Abstract

In this paper the authors present a structure analysis of the state space of two particular classes of variable structure systems, i.e. time-varying linear systems and constant bilinear systems.

They prove that in both cases a suitable choice of input-state and state-output interaction properties makes it possibile to effect a canonical decomposition of the state space. This decomposition preserves the essential characteristics of the well known canonical decomposition valid in the case of constant linear systems.

Introduction

Several classes of dynamical systems can be considered as "variable structure" systems, because of the lack of a precise definition for these latter. For example, the "bilinear" systems are usually regarded as variable structure systems. Actually, these systems may be considered as derived from constant linear systems in which some of the parameters

---

* Centro di Studio dei Sistemi di Controllo e Calcolo Automatici, C.N.R., Roma

** Istituto di Automatica, Università di Roma

*** Istituto di Automatica, Università di Roma e Fondazione V. Bordoni, Roma

This work was supported by C.N.R.

are varied linearly by means of a control action. Therefore, it would also seem reasonable to consider as "variable structure" systems those linear systems in which the parameters vary autonomously in time, i.e. the time-varying linear systems. These two particular classes of systems seem to be the simplest types of systems in which the structure is variable, the variation being autonomous in one case and forced in the other. In the first of these two cases the study is made more complex by the variability with time, but it can always be founded on the essential characteristic of linearity; in the second case non-linearity intervenes in the simplest manner that one can think of, and this makes it possible to base the analysis on the use of techniques that are derived from those employed in the linear case (as will be shown later on).

The present paper is concerned with the structure analysis of the state space of these two classes of systems. The analysis is developed on the basis of suitable input-state and state-output interaction properties, these properties being chosen in such a way as to ensure a decomposition with characteristics similar to those of the decomposition of fixed systems and, therefore, such as to allow the identification of the part that is uniquely connected with the input-output descriptions [1][2].

Canonical Decomposition of Time-Varying Linear Systems

The systems considered in this section are those described by equations of the type

$$\begin{aligned} \dot{x}(t) &= A(t)\, x(t) + B(t)\, u(t) \\ y(t) &= C(t)\, x(t) \end{aligned} \tag{1}$$

where A, B, C are $n\times n$, $n\times p$, and $q\times n$ matrices of continuous functions of t, $x(t) \varepsilon R$ denotes the state variable at time t, $u(t)\varepsilon R^p$ the input variable, and $y(t) \varepsilon R^q$ the output variable The input u is assumed to be continuous and hence the corresponding solution x is of class C^1 and y of class C°.

It is natural to associate with a given "system" the set of all the "descriptions" of type (1) obtained through a change of variables in the system state space, $z(t) = T(t)x(t)$, where T(t) is an $n\times n$ non-singular matrix of C^1

functions. Consequently, the set R of matrix triplets $\{A(\cdot), B(\cdot), C(\cdot)\}$ that corresponds to these descriptions constitutes an equivalence class with respect to the group T_a of transformations $\Theta : R \to R$ defined by the relation

$$T_a = \{\Theta : \Theta[A(t), B(t), C(t)] = [T(t)A(t)T^{-1}(t) + \dot{T}(t)T^{-1}(t), T(t)B(t), C(t)T^{-1}(t)]\} \tag{2}$$

As is well known, any two matrix triplets belonging to R, and likewise the corresponding descriptions of type (1), are said to be algebraically equivalent.

Within the class of equations defined in this way [3] it is usual to look for the existence of canonical forms of the equations that may correspond to the decomposition of the state space performed on the basis of the properties of controllability (reachability) and observability (or constructability). However, for time-varying systems the above properties do not lead to canonical forms unless restrictive hypotheses are imposed on the system [4]. In order to overcome this obstacle, two new structural properties (but based on the ones mentioned above) have been introduced: influenceability and invisibility. These properties give rise to a decomposition of the state space into subspaces with constant dimensions and continuously evolving in time; this guarantees the existence, in the general case, of canonical form for the equations.

The structure theory which can be developed on the basis of these properties is presented in [1]. Only the main results without proofs will be stated here.

As a starting point, the two new properties (influenceability and invisibility) for the system described by (1) are defined as follows:

Definition 1: A state x is influenceable at time t if it can be expressed as a sum of states, each of which is at least controllable or reachable at the same time t.

Definition 2: A state is invisible at time t if it is unobservable and unconstructable at time t.

As can readily be verified, the sets of the influenceable and invisible states are linear subspaces for every value of t; they will henceforth be denoted respectively by

$X_p(t)$ and $X_q(t)$. In order to identify these subspaces, it now becomes necessary to introduce the Gramian matrices

$$P(t;\eta,\zeta) = \int_{\eta}^{\zeta} \Phi(t,\tau)B(\tau)B^T(\tau)\Phi^T(t,\tau)d\tau \quad (3)$$

$$Q(t;\eta,\zeta) = \int_{\eta}^{\zeta} \Phi^T(\tau,t)C^T(\tau)C(\tau)\Phi(\tau,t)d\tau \quad (4)$$

where $\Phi(t,\tau)$ denotes the state transition matrix of the system (1).

Let $R[P]$ denote the range space of P, and $N[Q]$ the null space of Q; one can state the following two theorems:

Theorem 1: The subspace $X_p(t)$ of all the influenceable states at each time t can be expressed in the form

$$X_p(t) = R[P(t;\hat{\eta},\hat{\zeta})] \quad (5)$$

where $\hat{\eta},\hat{\zeta}$ is any pair of values of η,ζ for which, at an arbitrarily fixed time t_o, $P(t_o;\eta,\zeta)$ has maximum rank.

Furthermore,

$$\dim X_p(t) = \text{constant} \quad (6)$$

Theorem 2: The subspace $X_q(t)$ of all the invisible states at each time t can be expressed in the form

$$X_q(t) = N[Q(t;\bar{\eta},\bar{\zeta})] \quad (7)$$

where $\bar{\eta},\bar{\zeta}$ is any pair of values of η,ζ for which, at an arbitrarily fixed time t_o, $Q(t_o;\eta,\zeta)$ has maximum rank.

Furthermore,

$$\dim X_q(t) = \text{constant} \quad (8)$$

With reference to the two subspaces considered above, it becomes possible to effect the decomposition of the state space X of the system (1) at each time t by means of the relations introduced by R.E. Kalman in [5], i.e.

$$A(t) = X_p(t) \cap X_q(t) \quad (9)$$

$$\begin{aligned} \mathcal{X}_p(t) &= \mathcal{A}(t) \oplus \mathcal{B}(t) \\ \mathcal{X}_q(t) &= \mathcal{A}(t) \oplus \mathcal{C}(t) \\ \mathcal{X} &= \mathcal{A}(t) \oplus \mathcal{B}(t) \oplus \mathcal{C}(t) \oplus \mathcal{D}(t) \end{aligned} \tag{9}$$

It follows from Theorems 1 and 2 that the four subspaces identified in this way have constant dimensions. Moreover, it is possible to prove that they can be chosen in such a way as to evolve continuously in time. One can therefore state the following

Theorem 3 (Main Theorem): Taking as basis of the state space the union of suitable bases in the four subspaces $\mathcal{A}(.)$, $\mathcal{B}(.)$, $\mathcal{C}(.)$ and $\mathcal{D}(.)$, equations (1) assume the form

$$\begin{bmatrix} \dot{x}_a(t) \\ \dot{x}_b(t) \\ \dot{x}_c(t) \\ \dot{x}_d(t) \end{bmatrix} = \begin{bmatrix} A_{aa}(t) & A_{ab}(t) & A_{ac}(t) & A_{ad}(t) \\ 0 & A_{bb}(t) & 0 & A_{bd}(t) \\ 0 & 0 & A_{cc}(t) & A_{cd}(t) \\ 0 & 0 & 0 & A_{dd}(t) \end{bmatrix} \begin{bmatrix} x_a(t) \\ x_b(t) \\ x_c(t) \\ x_d(t) \end{bmatrix} + \begin{bmatrix} B_a(t) \\ B_b(t) \\ 0 \\ 0 \end{bmatrix} u(t) \tag{10}$$

$$y(t) = (0 \quad C_b(t) \quad 0 \quad C_d(t)) \begin{bmatrix} x_a(t) \\ x_b(t) \\ x_c(t) \\ x_d(t) \end{bmatrix}$$

where $[x_a^T(t)\ \ 0\ \ 0\ \ 0]^T$, $[0\ \ x_b^T(t)\ \ 0\ \ 0]^T$, $[0\ \ 0\ \ x_c^T(t)\ \ 0]^T$, and $[0\ \ 0\ \ 0\ \ x_d^T(t)]^T$, are coordinates of vector belonging to $\mathcal{A}(t)$, $\mathcal{B}(t)$, $\mathcal{C}(t)$ and $\mathcal{D}(t)$ respectively, and the coefficient matrices are continuous (canonical form).

The triplets of matrices that assume the form associated with equation (10) are an equivalence class with respect to the subgroup of $\mathcal{T}_a$ defined by

$$\mathcal{T}_{ac} = \{\theta \varepsilon \mathcal{T}_a;\ T(t) = \begin{bmatrix} T_{aa}(t) & T_{ab}(t) & T_{ac}(t) & T_{ad}(t) \\ 0 & T_{bb}(t) & 0 & T_{bd}(t) \\ 0 & 0 & T_{cc}(t) & T_{cd}(t) \\ 0 & 0 & 0 & T_{dd}(t) \end{bmatrix}\} \tag{11}$$

where the matrices on the main diagonal are square and have the same dimension of the subspaces $\mathcal{A}(.)$, $\mathcal{B}(.)$, $\mathcal{C}(.)$ and $\mathcal{D}(.)$ respectively.

Apart from the intrinsic interest related to the theory that has just been presented, it is immediately apparent that it makes it possible to investigate the connection with the problem of the minimal realization of a given weighting pattern. In fact, one can at once verify that this latter depends only on the part "b" of the decomposition. On the other hand, it is also possible to prove, making use of the theory of realization developed by D.C. Youla [6], that a realization is minimal if and only if its state space possesses the properties that characterize the part "b", i.e. if it is "completely influenceable" and "completely visible".

Canonical Decomposition of Constant Bilinear Systems

The systems considered in this section are those described by equations of the type

$$\begin{aligned} \dot{x}(t) &= A\,x(t) + N\,x(t)u(t) + B\,u(t) \\ y(t) &= C\,x(t) \end{aligned} \tag{12}$$

where $u(t) \in R$ is the input, $y(t) \in R$ is the output, and $x(t) \in R^n$ is the state at time t. The matrices A, N, B, C are constant matrices of suitable dimensions.

In this case, too, it is natural to associate with a given "system" the set of all the "descriptions" of type (2) obtained through a change of variables in the system state space, $z(t) = T x(t)$, where T is an n×n non-singular constant matrix. Consequently, the group of transformations with respect to which the quadruplets {A, N, B, C} associated with these descriptions constitute an equivalence class is defined by

$$\mathcal{T} = \{\theta:\theta[A,N,B,C] = [TAT^{-1},TNT^{-1},TB,CT^{-1}]\} \qquad (13)$$

In the case of the bilinear systems, once again, one can try to perform the analysis of the state space on the basis of input-state and state-output properties. Such an analysis has been developed by the authors in [2], considering the properties of reachability from the origin and unobservability and embedding the set of states having the former property in the least linear subspace (since this set does not itself constitute a subspace). For the proofs the reader is once again referred to the original paper and only the essential results will be mentioned here.

The two properties taken into consideration as the basis of the analysis are formalized in the following definitions:

Definition 3: A state x of the system (12) is said to be reachable from the origin if there exists an admissible input function that transfers the origin of the state space into the state x in a finite interval of time.

Definition 4: A state x of the system (12) is said to be unobservable if the difference between the output and the output from the zero state is identically zero for every admissible input function.

The set of states reachable from the origin does not constitute a linear space; as already mentioned above, for the purposes of identifying a canonical decomposition of the state space it is therefore necessary to consider the small est linear subspace that contains this set. This subspace will henceforth be denoted by X_p. The set of unobservable states, on the other hand, constitutes a linear subspace and will henceforth be represented by X_q.

In order to identify the subspaces, one now has to introduce the Gramian matrices (*)

$$\mathcal{P}_n\left[e^{At}N, e^{At}B\right] = \int_0^1 P_n(t_1,\ldots,t_n)P_n^T(t_1,\ldots,t_n)dt_1\ldots dt_n \tag{14}$$

$$\mathcal{Q}_n\left[C\,e^{At}, N\,e^{At}\right] = \int_0^1 Q_n^T(t_1,\ldots,t_n)Q_n(t_1,\ldots,t_n)dt_1\ldots dt_n \tag{15}$$

of the functions

$$P_n(t_1,\ldots,t_n) = \left[e^{At_1}B \,\vdots\, e^{At_2}Ne^{At_1}B \,\vdots\, \ldots \,\vdots\, e^{At_n}N\ldots e^{At_2}Ne^{At_1}B\right] \tag{14'}$$

$$Q_n(t_1,\ldots,t_n) = \begin{bmatrix} C\,e^{At_1} \\ C\,e^{At_1}N\,e^{At_2} \\ \vdots \\ C\,e^{At_1}N\,e^{At_2}\ldots N\,e^{At_n} \end{bmatrix} \tag{15'}$$

At this point one can state the following theorems:

Theorem 4: The least linear subspace that contains all the states of the system (12) reachable from the origin is

$$X_p = \mathcal{R}\{\mathcal{P}_n\left[e^{At}N, e^{At}B\right]\} \tag{16}$$

(*) The symbols adopted for the l.h.s. of (14) and (15) are the same as the one used in paper [2], where it was chosen with a view to meeting more general requirements; for the sake of consistency, it has not been simplified for the more limited purposes of the present paper.

Theorem 5: The subspace of the unobservable states of the system (12) is given by

$$\mathcal{X}_q = \mathcal{N}\{\mathcal{Q}_n[C\,e^{At}, N\,e^{At}]\} \tag{17}$$

Referring to the subspaces introduced in this way, and once again making use of the canonical decomposition of the state space identified by means of equations (9), one can readily arrive at stating the following

Theorem 6 (Main Theorem): Assuming as basis in the state space the union of bases in the four subspaces $\mathcal{A}$, $\mathcal{B}$, $\mathcal{C}$ and $\mathcal{D}$ of a canonical decompcsition, the equations (12) assume the form

$$\begin{bmatrix} \dot{x}_a(t) \\ \dot{x}_b(t) \\ \dot{x}_c(t) \\ \dot{x}_d(t) \end{bmatrix} = \begin{bmatrix} A_{aa} & A_{ab} & A_{ac} & A_{ad} \\ 0 & A_{bb} & 0 & A_{bd} \\ 0 & 0 & A_{cc} & A_{cd} \\ 0 & 0 & 0 & A_{dd} \end{bmatrix} \begin{bmatrix} x_a(t) \\ x_b(t) \\ x_c(t) \\ x_d(t) \end{bmatrix} + \begin{bmatrix} N_{aa} & N_{ab} & N_{ac} & N_{ad} \\ 0 & N_{bb} & 0 & N_{bd} \\ 0 & 0 & N_{cc} & N_{cd} \\ 0 & 0 & 0 & N_{dd} \end{bmatrix} \begin{bmatrix} x_a(t) \\ x_b(t) \\ x_c(t) \\ x_d(t) \end{bmatrix} u(t) + \begin{bmatrix} B_a \\ B_b \\ 0 \\ 0 \end{bmatrix} u(t)$$

$$y(t) = (0 \quad C_b \quad 0 \quad C_d) \begin{bmatrix} x_a(t) \\ x_b(t) \\ x_c(t) \\ x_d(t) \end{bmatrix} \tag{18}$$

where $(x_a^T \; 0 \; 0 \; 0)^T$, $(0 \; x_b^T \; 0 \; 0)^T$, $(0 \; 0 \; x_c^T \; 0)^T$ and

$(0 \quad 0 \quad 0 \quad x_d^T)^T$ are coordinates of vectors belonging respectively to the subspaces $\mathcal{A}$, $\mathcal{B}$, $\mathcal{C}$ and $\mathcal{D}$ (canonical form).

The quadruplets {A, N, B, C}, which assume the form associated with (18), are an equivalence class with respect to the subgroup of $\mathcal{T}$ defined by

$$\mathcal{T}_c = \{\theta \varepsilon \mathcal{T}, T = \begin{bmatrix} T_{aa} & T_{ab} & T_{ac} & T_{ad} \\ 0 & T_{bb} & 0 & T_{bd} \\ 0 & 0 & T_{cc} & T_{cd} \\ 0 & 0 & 0 & T_{dd} \end{bmatrix} \} \qquad (19)$$

where the matrices on the main diagonal are square and have the same dimension of the subspaces $\mathcal{A}$, $\mathcal{B}$, $\mathcal{C}$, $\mathcal{D}$ respectively.

In connection with the canonical form that has just been introduced, and which is characterized by a direct analogy with the results obtained in the case of the linear systems, one can readily verify that the kernels of the Volterra series expansion of the zero state response of the system (12) depend only on the matrices associated with the part "b" of the decomposition. But what is even more important to note is the fact that, on the basis of the realization theory developed by the authors in [2], a realization of a given sequence of kernels of the Volterra series expansion of the zero state response is minimal if and only if its state space is observable and spanned by the states reachable from the origin.

Conclusions

It is interesting to note that, by considering simple but important classes of variable structure systems, it is possible to develop a structure analysis that preserves the essential characteristics of the structure analysis peculiar to fixed structure systems (linear and constant). It is also interesting to stress that these results can be obtained by basing substantially on the methodologies characteristic of the analysis of the linear systems.

REFERENCES

1 P.d'Alessandro, A. Isidori, A. Ruberti, A new approach to the theory of canonical decomposition of linear dynamical systems, Report 1-10, Istituto di Automatica, Università di Roma. To appear on SIAM J. on Control (1973).

2 P.d'Alessandro, A. Isidori, A. Ruberti, Realization and structure theory of bilinear dynamical systems. Reports 2-04 and 2-16, Istituto di Automatica, Università di Roma. To be published.

3 L. Weiss, R.E. Kalman, Contribution to linear system theory, Internat. J. Engrg. Sci. 3 (1965), pp. 141-171.

4 L. Weiss, On the structure theory of linear differential systems, SIAM J. on Control, (1968), pp. 659-680.

5 R.E. Kalman, Canonical structure of linear dynamical systems, Proc. Nat. Acad. Sci. U.S.A., 48 (1962) pp. 596-600.

6 D.C. Youla, The synthesis of linear dynamical systems from prescribed weighting pattern, SIAM J. Appl. Math. 14 (1966), pp. 527-549.

STRUCTURAL PROBLEMS OF REGULATORS*

G.Basile and G.Marro

University of Genoa, Italy and
University of Bologna, Italy

ABSTRACT

The structure of linear dynamical systems is analized by means of recursive matrix operations in order to test the feasibility of particular types of controls. Structural requirements for achieving such properties as decoupling, disturbance insensitivity and parametric insensitivity are stated in mathematical terms and effective synthesis procedures for multivariable special purpose dynamic regulators are illustrated.

INTRODUCTION

In this paper some previously introduced concepts and algorithms [1-5] are used in order to characterize in a simple geometric way some structural properties of linear dynamical plants in connection with the problem of synthesizing dynamic regulators which achieve particular features of the overall system.

It is assumed that the definitions and basic properties of controlled and conditioned invariance and of constrained controllability are known. However for the readers convenience a short review of them is reported in the Appendix.

Lowercase boldface letters ($\mathbf{a},\mathbf{b}$) denote vectors, capital boldface letters ($\mathbf{A},\mathbf{B}$) linear transformations or matrices, capital script ($\mathcal{A},\mathcal{B}$)subspaces. By $\mathbf{x} \in R^n$ is meant that $\mathbf{x}$ is a n-vector, by $\mathcal{X} \subseteq R^n$ is meant that $\mathcal{X}$ is a set of n-vectors. The symbols $\mathbf{e}_1, \mathbf{e}_2, \ldots, \mathbf{e}_n$ are used to denote the

* This research has been supported by CNR (National Research Council), Rome, Italy.

reference basis vectors, which in R^n are the column vectors of the $n \times n$ identity matrix. $\mathcal{R}(\mathbf{A})$, $\mathcal{N}(\mathbf{A})$ denote the range and the null-space of the linear transformations $\mathbf{A}$. $\mathbf{A}\mathcal{X}$ is the image of the set $\mathcal{X}$ under the linear transformation $\mathbf{A}$, $\mathbf{A}^{-1*}\mathcal{Y}$ is the inverse image of $\mathcal{Y}$ under the linear transformation $\mathbf{A}$, i.e. the locus of vectors which are mapped into $\mathcal{Y}$ by $\mathbf{A}$. $\mathcal{X}^{\perp}$ is the orthogonal complement of $\mathcal{X}$, $\mathrm{mi}(\mathbf{A},\mathcal{X})$ is the minimum $\mathbf{A}$-invariant containing $\mathcal{X}$, $\mathrm{MCI}(\mathbf{A},\mathcal{B},\mathcal{X})$ is the maximum $(\mathbf{A},\mathcal{B})$-controlled invariant contained in $\mathcal{X}$, $\mathrm{MRS}(\mathbf{A},\mathcal{R}(\mathbf{B}),\mathcal{X})$ is the maximum reachable subspace contained in $\mathcal{X}$ (see the Appendix).

Consider the controlled system represented as a block in Fig.1,a. The vectors $\mathbf{x}\in R^n$, $\mathbf{u}\in R^m$, $\mathbf{d}\in R^r$ and $\mathbf{y}\in R^s$ are respectively the state, the manipulated input, the non-manipulated input and the output. The non-manipulated input is called "disturbance" in the sequel and is assumed to be accessible (directly or not) for measurement.

Let the mathematical model of the controlled system be

$$\dot{\mathbf{x}}(t) = \mathbf{A}\,\mathbf{x}(t) + \mathbf{B}\,\mathbf{u}(t) + \mathbf{D}\,\mathbf{d}(t) \tag{1a}$$

$$\mathbf{y}(t) = \mathbf{C}\,\mathbf{x}(t) \tag{1b}$$

This model is assumed to be completely observable and completely controllable by the manipulated input, the matrices $\mathbf{A}$, $\mathbf{B}$, $\mathbf{C}$ and $\mathbf{D}$ are constant unless otherwise stated.

If a linear dynamic[1] controller is connected to the plant as shown in Fig.1,b, an overall system is obtained which is described by a set of equations analogous to (1), namely

$$\dot{\hat{\mathbf{x}}}(t) = \hat{\mathbf{A}}\,\hat{\mathbf{x}}(t) + \hat{\mathbf{B}}\,\mathbf{v}(t) + \hat{\mathbf{D}}\,\mathbf{d}(t) \tag{2a}$$

$$\mathbf{y}(t) = \hat{\mathbf{C}}\,\hat{\mathbf{x}}(t) \tag{2b}$$

where the augmented state $\hat{\mathbf{x}}$ includes both the system and regulator states and $\mathbf{v}\in R^h$ is the input to the regulator.

STRUCTURAL ANALYSIS AND INVARIANT SUBSPACES

Usual requirements for the overall system performance, such as decoupling, disturbance insensitivity and parameter

[1] Not necessarily purely dynamic.

insensitivity will be expressed by inclusion relationships of characteristic subspaces of the linear transformation which appear in equations (2), as $\hat{\mathbf{A}}$-invariants and ranges and null-spaces of $\hat{\mathbf{B}}$, $\hat{\mathbf{C}}$ and $\hat{\mathbf{D}}$. Then it will be shown that, in order to obtain such requirements, the controlled system must present specific corresponding structural features, which can be expressed in a similar way in terms of $(\mathbf{A}, \mathscr{R}(\mathbf{B}))$-controlled invariants and ranges and null-spaces of $\mathbf{C}$ and $\mathbf{D}$.

a) Decoupling or non-interaction

Definition 1: The overall system (2) is non-interacting with respect to the couples of input and output subspaces $\mathscr{V}_i , \mathscr{Y}_i'$ (i=1,...,k') if

$$\mathscr{Y}_i' = \hat{\mathbf{C}} \operatorname{mi}(\hat{\mathbf{A}}, \hat{\mathbf{B}} \mathscr{V}_i) \qquad (i=1,\dots,k') . \tag{3}$$

Usuall both $\mathscr{V}_i$ and $\mathscr{Y}_i'$ correspond to given subsets of inputs and outputs, so that they are coordinate subspaces and can be defined as

$$\mathscr{V}_i = \operatorname{span}(\mathbf{e}_j \in R^h , \; j \in I_i^v) \qquad (i=1,\dots,k') \tag{4}$$

$$\mathscr{Y}_i' = \operatorname{span}(\mathbf{e}_j \in R^s , \; j \in I_i^{y'}) \qquad (i=1,\dots,k') , \tag{5}$$

being I_i^v and $I_i^{y'}$ (i=1,...,k') given subsets of the sets of integers $\{1,\dots,h\}$ and $\{1,\dots,s\}$.

b) Disturbance insensitivity

Definition 2: The overall system (2) is disturbance insensitive with respect to the couples of disturbance and output subspaces $\mathscr{D}_i$, $\mathscr{Y}_i''$(i=1,...,k") if

$$\mathscr{Y}_i''^{\perp} \supseteq \hat{\mathbf{C}} \operatorname{mi}(\hat{\mathbf{A}}, \hat{\mathbf{D}} \mathscr{D}_i) \qquad (i=1,\dots,k'') \tag{6}$$

Also in this case the input and output subspaces usually are coordinate subspaces corresponding to subsets of indexes I_i^d, $I_i^{y''}$(i=1,...,k").

c) Parametric insensitivity

Parametric insensitivity is approached together with

decoupling and disturbance insensitivity by extending the previous Definition 1 and 2 to an augmented system. Suppose that a parameter vector $\mathbf{p} \in R^q$ affects the system behaviour, so that equations (1) become

$$\dot{\mathbf{x}}(t) = \mathbf{A}(\mathbf{p})\,\mathbf{x}(t) + \mathbf{B}(\mathbf{p})\,\mathbf{u}(t) + \mathbf{D}(\mathbf{p})\,\mathbf{d}(t) \tag{7a}$$

$$\mathbf{y}(t) = \mathbf{C}(\mathbf{p})\,\mathbf{x}(t) \quad . \tag{7b}$$

Consider the set of partial derivatives

$$\mathbf{s}_i^x(t) = \left.\frac{\partial\,\mathbf{x}(t,\mathbf{p})}{\partial p_i}\right]_{\mathbf{p}=\mathbf{p}^*} \qquad \mathbf{s}_i^y(t) = \left.\frac{\partial\,\mathbf{y}(t,\mathbf{p})}{\partial p_i}\right]_{\mathbf{p}=\mathbf{p}^*} \qquad (i=1,..,q) \tag{8}$$

which are called the state and output sensitivity functions in the neighbourhood of $\mathbf{p}^*$. They satisfy the linear differential system

$$\dot{\mathbf{x}}(t) = \mathbf{A}(\mathbf{p}^*)\mathbf{x}(t) + \mathbf{B}(\mathbf{p}^*)\mathbf{u}(t) + \mathbf{D}(\mathbf{p}^*)\mathbf{d}(t) \tag{9a}$$

$$\dot{\mathbf{s}}_i^x(t) = \mathbf{A}_i\,(t) + \mathbf{A}(\mathbf{p}^*)\,\mathbf{s}_i^x(t) + \mathbf{B}_i\mathbf{u}(t) + \mathbf{D}_i\mathbf{d}(t) \qquad (i=1,..,q) \tag{9b}$$

$$\mathbf{y}(t) = \mathbf{C}(\mathbf{p}^*)\mathbf{x}(t) \tag{9c}$$

$$\mathbf{s}_i^y(t) = \mathbf{C}_i\mathbf{x}(t) + \mathbf{C}(\mathbf{p}^*)\,\mathbf{s}_i^x(t) \qquad (i=1,..,q) \tag{9d}$$

where

$$\mathbf{A}_i = \left.\frac{\partial \mathbf{A}}{\partial p_i}\right]_{\mathbf{p}=\mathbf{p}^*}, \quad \mathbf{B}_i = \left.\frac{\partial \mathbf{B}}{\partial p_i}\right]_{\mathbf{p}=\mathbf{p}^*}$$

$$\mathbf{C}_i = \left.\frac{\partial \mathbf{C}}{\partial p_i}\right]_{\mathbf{p}=\mathbf{p}^*} \quad \mathbf{D}_i = \left.\frac{\partial \mathbf{D}}{\partial p_i}\right]_{\mathbf{p}=\mathbf{p}^*}$$

If the initial state is assumed to be insensitive, the initial conditions for the insensitivity functions are $\mathbf{s}_i^x(0) = \mathbf{o}$ $(i=1,...,q)$. A state or output trajectory is said to be insensitive to the parameter p_i if $\mathbf{s}_i^x(t) = \mathbf{o}$ or $\mathbf{s}_i^y(t) = \mathbf{o}$, $t \geqslant 0$. An insensitive system is one which evolves along insensitive trajectories

Equations (9) provide sensitivity functions besides state and output trajectories and are of the same type as equations (1). Of course, in order to deduce the sensitivity functions of the overall system a similar procedure can be applied, because the matrices $\hat{\mathbf{A}}, \hat{\mathbf{B}}, \hat{\mathbf{C}}$ and $\hat{\mathbf{D}}$ of equations (2) are indirectly functions of the parameter $\mathbf{p}$ as well.

Clearly, when decoupling or disturbance insensitivity

are considered together with insensitivity to given sets of parameters $S_i=\{p_j : j \in I_i^p\}$ (i=1,...,k' or k"), where I_i^p are s subsets of indexes, conditions of the same type as (3) or (6) can be stated, but referring to an augmented system whose state and output include sensitivity funtions. In fact insensitivity prerequisites are equivalent to restrict the zero-state response of the augmented system to range over given subspaces.

STRUCTURAL CONDITIONS IN TERMS OF CONTROLLED INVARIANTS AND REACHABLE SUBSPACES

In the previous section it has been shown that two basic structural properties of linear multivariable systems, decoupling and disturbance insensitivity, are expressed by conditions of the same type. Conditions (3) are necessary and sufficient in order that the zero-state response to any input ranging on $\mathscr{V}_i$ ranges on $\mathscr{Y}_i'$ and every point of $\mathscr{Y}_i'$ can be reached by means of a suitable input ranging on $\mathscr{V}_i$. Conditions (6) are necessary and sufficient in order that zero-state response to any disturbance input ranging on $\mathscr{D}_i$ ranges on $\mathscr{Y}_i''^{\perp}$.

The feasibility of (3) and (6) depends, of course, on the structure of the controlled system. Necessary conditions can be stated in terms of maximum controlled invariants and reachable subspaces and checked by means of Algorithms 3 and 5 reported in Appendix.

Assertion 1: A dynamic control apparatus which satisfies relationships (3) exixts only if

$$\mathscr{Y}_i' = \mathbf{C}\,\mathrm{MRS}(\mathbf{A},\mathscr{R}(\mathbf{B}),\mathbf{C}^{-1*}\mathscr{Y}_i') \quad (i=1,..,k') . \qquad (10)$$

Conditions (10) are a straightforward application of constrained controllability. In fact, for the controlled system to be controllable along zero-state output trajectories ranging on $\mathscr{Y}_i'$ it is clearly necessary that the state evolves on the maximum reachable subspace which maps into $\mathscr{Y}_i'$. If one of (10) is not satisfied, i.e.

$$\mathscr{Y}_i' \supset \mathbf{C}\,\mathrm{MRS}(\mathbf{A},\mathscr{R}(\mathbf{B}),\ \mathbf{C}^{-1*}\mathscr{Y}_i') , \qquad (11)$$

it follows that decoupling is not compatible with complete controllability. Indeed not all point of $\mathscr{Y}_i'$ are reachable by output trajectories ranging on $\mathscr{Y}_i'$, but only the points of the subspace in the right side member of (11)

Assertion 2: A dynamic control apparatus which satisfies relationships (6) exists only if

$$\mathbf{D}\mathcal{D}_i \subseteq \mathrm{MCI}(\mathbf{A},\mathcal{R}(\mathbf{B}),\mathbf{C}^{-1*}\mathcal{Y}_i''^{\perp})+\mathcal{R}(\mathbf{B}) \quad (i=1,..,k'') \qquad (12)$$

Conditions (12) mean that, in order to make any disturbance action ranging on $\mathcal{D}_i$ to correspond to a zero-state, output ranging $\mathcal{Y}_i''^{\perp}$, a control must exist such that the combined effects on state velocity of manipulated and non-manipulated inputs belong to the maximum controlled invariant which maps in $\mathcal{Y}_i''^{\perp}$.

Conditions (10) and (12) correspond to easily implementable matrix operations which allow to check if the structure of the controlled system is compatible with decoupling and disturbance insensitivity.

SYNTHESIS OF DECOUPLING AND DISTURBANCE INSENSITIVE CONTROLLERS

Now it will be shown that, if conditions stated in Assertions 1 and 2 are satisfied, it is possible to synthesize a dynamic controller which achieves decoupling and disturbance insensitivity. In the case of decoupling the controller allows also the arbitrary pole assignment, while in the case of disturbance insensitivity all poles can be arbitrarily assigned only if a further condition is satisfied.

The controller is implementable according to the block diagram shown in Fig.2. Note that this is essentially a feedforward implementation: indeed, as far as structural properties are concerned, feedback does not give any improvement.

Arbitrary pole assignment for the controlled system, which is assumed to be completely controllable and observable, can be achieved by means of a suitable dynamic compensator obtained by combining a state observer with an algebraic feedback [6-9]. Such a stabilizing device, shown by dotted lines in Fig.2, is actually a part of the controller; nevertheless, as far as the further mathematical developments are concerned, it will be considered as a part of the controlled system and the matrices **A** , **B**, **C** and **D** will be properly redefined.

The feedforward section of the controller is composed of a certain number of dynamic units, one for each couple of non-interacting input-output and insensitive disturbance-out put subspaces.

Each of the decoupling units has the structure represented in Fig.3. Suppose that the i-th of relationships (10) is satisfied. The purpose of the unit is to generate control signals corresponding to state trajectories ranging on the reachable subspace which appears in the right side member of (10), namely

$$\mathscr{I}_i' = \mathrm{MRS}(\mathbf{A}, \mathscr{R}(\mathbf{B}), \mathbf{C}^{-1*}\mathscr{Y}_i') . \qquad (13)$$

Each unit contains a model of the controlled system described by the differential equations.

$$\dot{\mathbf{x}}_i'(t) = \mathbf{A}\mathbf{x}_i'(t) + \mathbf{B}\mathbf{u}_i(t) \quad . \qquad (14)$$

According to Property 2 of the Appendix, it is possible to determine a feedback matrix $\mathbf{H}_i'$ such that the controlled invariant (13) is transformed into a simple invariant under $\mathbf{A}+\mathbf{B}\mathbf{H}_i'$. Because of Property 5, it is not necessary to build a complete model of the controlled system, being necessary only the state coordinates corresponding to a basis of $\mathscr{I}_i'$.

The feedforward block of the unit is described by the equation

$$\mathbf{w}_i' = \mathbf{G}_i'\mathbf{v} \quad , \qquad (15)$$

where the matrix $\mathbf{G}_i'$ has maximal rank and is chosen in such a way that

$$\mathbf{B}\mathbf{G}_i'\mathscr{V}_i \subseteq \mathscr{I}_i' \quad . \qquad (16)$$

Hence, starting from the zero-state and by varying anyway the input $\mathbf{v}$ to the controller on the subspace $\mathscr{V}_i$, only motions on $\mathscr{I}_i'$ are possible both for the model and the controlled system, which are subject to the same control. In force of Property 5, being $\mathscr{I}_i'$ a reachable controlled invariant, the eigenvalues of each unit can be arbitrarily assigned by a proper choice of $\mathbf{H}_i'$.

Up to this point it has been shown that under the necessary conditions stated in Assertion 1 a decoupling regulator with arbitrary eigenvalues can be synthesized.

Let us consider now the disturbance compensating units, which are realized as shown in Fig.4. The philosophy according to which they are designed is substantially analogous to that of decoupling units. The only difference is that the inputs whose effect has to be constrained to given output subspaces are not manipulated.

Suppose that relationships (12) are satisfied and let

$$\mathcal{X}_i = \mathrm{MCI}(\mathbf{A}, \mathcal{R}(\mathbf{B}), \mathbf{C}^{-1*}\mathcal{Y}_i''^{\perp}) \tag{17}$$

be the controlled invariant appearing at the right side member of the i-th of them. Of course, $\mathcal{X}_i$ can be transformed into a simple $(\mathbf{A}+\mathbf{BH}_i'')$-invariant by means of a proper state feedback $\mathbf{H}_i''$, but stability is not assured, because $\mathcal{X}_i$ in general is not completely reachable by acting on the manipulated input.

The reachable part of $\mathcal{X}_i$ is

$$\mathcal{W}_{1,i} = \mathrm{mi}(\mathbf{A}+\mathbf{BH}_i'', \; \mathcal{X}_i \cap \mathcal{R}(\mathbf{B})) \quad , \tag{18}$$

and only its associated eigenvalues can be arbitrarily assigned by a proper choice of $\mathbf{H}_i''$.

According to relationships (12), for each unit a feedforward matrix $\mathbf{G}_i''$ can be determined such that

$$(\mathbf{BG}_i'' + \mathbf{D})\mathcal{D}_i \subseteq \mathcal{X}_i \quad . \tag{19}$$

In this way the disturbance inputs ranging on $\mathcal{D}_i$ can be compensated being the corresponding state velocity variations on the controlled invariant $\mathcal{X}_i$.

The set of states which are reachable on $\mathcal{X}_i$ by the action of the compensated disturbances is

$$\mathcal{W}_{2,i} = \mathrm{mi}(\mathbf{A}+\mathbf{BH}_i'', \; (\mathbf{BG}_i'' + \mathbf{D})\mathcal{D}_i) \quad , \tag{20}$$

a $(\mathbf{A}, \mathcal{R}(\mathbf{B}))$-controlled invariant too.

The model contained in the i-th disturbance compensating unit is described by the differential equations

$$\dot{\mathbf{x}}_i''(t) = \mathbf{A}\mathbf{x}_i''(t) + \mathbf{B}\mathbf{u}_i(t) + \mathbf{D}\mathbf{d}(t) \tag{21}$$

and reproduces the $(\mathbf{A}+\mathbf{BH}_i'')$-invariant

$$\mathcal{J}_i'' = \mathcal{W}_{1,i} + \mathcal{W}_{2,i} = \mathrm{mi}(\mathbf{A}+\mathbf{BH}_i'', (\mathbf{BG}_i''+\mathbf{D})\mathcal{D}_i + \mathcal{X}_i \cap \mathcal{R}(\mathbf{B})) . \tag{22}$$

Note that $\mathcal{J}_i''$ does not depend on the particular choice of $\mathbf{G}_i''$ because by varying $\mathbf{G}_i''$ while preserving the validity of relationships (19) subspaces $\mathcal{W}_{2,i}$ are obtained which differ each other by vectors belonging to $\mathscr{W}_{1,i}$. In force of Property 5 the model is stabilizable by properly choosing the matrix $\mathbf{H}_i''$ if, and only if, all associated eigenvalues with non-negative real parts are also associated to $\mathscr{W}_{1,i}$.

Thus necessary and sufficient conditions for the possibility of neutralizing disturbances by means of stable compensating units have been stated. They complete an early approach of the authors to this problem [10]. The minimization of the order of the units is an interesting still unsolved related problem.

CONCLUSIONS

The object of this note has been to present a convenient means for characterizing some important structural features of linear systems and to describe a systematic procedure for realizing special purpose controllers which achieve, whenever possible, any specified behaviour concerning input-output decoupling, disturbance insensitivity and parametric insensitivity.

Stability being the most important requirement for controllers, necessary and sufficient conditions for obtaining the prescribed structural properties together with stability or pole assigment have been discussed.

In order to emphasize the unitary essence of the approach and to explain its advantages in the simplest way, a quite criticizable assumption has been introduced, namely that all non-manipulated inputs are accessible for a continuous measurement. Of course, that appens very seldom in practice, so that, in order to compensate completely unknown disturbance inputs it is necessary to resort to particular techniques of unknown-input state observation or system inversion, like those discussed by the authors employing again the concepts of controlled and conditioned invariance [10-12].

APPENDIX

A. Simple, controlled and conditioned invariants.

Given a linear map $\mathbf{A}$ from R^n to R^n, a subspace $\mathscr{I} \subseteq R^n$ is said to be a $\mathbf{A}$-invariant if $\mathbf{A}\mathscr{I} \subseteq \mathscr{I}$. The sum and the intersection of two or more $\mathbf{A}$-invariants are clearly $\mathbf{A}$-invariants.

The sum of all $\mathbf{A}$-invariants contained in a given subspace $\mathscr{X} \subseteq R^n$ is called the maximum $\mathbf{A}$-invariant contained in $\mathscr{X}$, the intersection of all **A**-invariants containing a given subspace $\mathcal{X} \subseteq R^n$ is called the minimum **A**-invariant containing $\mathcal{X}$. They are provided by the following algorithms, which correspond to easily implementable matrix operations.

Algorithm 1: The maximum **A**-invariant contained in a given subspace $\mathcal{X}$, which is shortly denoted by MI($\mathbf{A}$, $\mathcal{X}$), is provided by the following sequence of subspaces, which converges at most in n-1 steps.

$$\mathcal{Z}_0 = \mathcal{X}$$

$$\mathcal{Z}_i = \mathcal{X} \cap \mathbf{A}^{-1*} \mathcal{Z}_{i-1} \qquad (i=1,2,..) .$$

Algorithm 2: The minimum **A**-invariant containing a given subspace $\mathcal{X}$, which is shortly denoted by mi($\mathbf{A}$, $\mathcal{X}$), is provided by the following sequence of subspaces, which converges at most in n-1 steps.

$$\mathcal{Z}_0 = \mathcal{X}$$

$$\mathcal{Z}_i = \mathcal{X} + \mathbf{A}\mathcal{Z}_{i-1} \qquad (i=1,2,..) .$$

In order to provide a simple geometrical approach to many problems of multivariable linear system theory, controlled and conditioned invariants have been introduced as extensions of simple invariants.

Given a linear map **A** from R^n to R^n and a subspace $\mathcal{B} \subseteq R^n$, a subspace $\mathcal{I} \subseteq R^n$ is said a ($\mathbf{A}$, $\mathcal{B}$)-controlled invariant if $\mathbf{A}\mathcal{I} \subseteq \mathcal{I} + \mathcal{B}$. The sum of two or more ($\mathbf{A}$, $\mathcal{B}$)-controlled invariants is a ($\mathbf{A}$, $\mathcal{B}$)-controlled invariant, so that the maximum controlled invariant contained in a given subspace $\mathcal{X} \subseteq R^n$ is univocally defined.

Given a linear map **A** from R^n to R^n and a subspace $\mathcal{B} \subseteq R^n$, a subspace $\mathcal{I} \subseteq R^n$ is said a ($\mathbf{A}$, $\mathcal{B}$)-conditioned invariant if $\mathbf{A}(\mathcal{I} \cap \mathcal{B}) \subseteq \mathcal{I}$. The intersection of two or more ($\mathbf{A}$, $\mathcal{B}$)-conditioned invariants is a ($\mathbf{A}$, $\mathcal{B}$)-conditioned invariant, so that the minimum conditioned invariant containing a given subspace $\mathcal{X} \subseteq R^n$ is univocally defined.

Algorithm 3: The maximum ($\mathbf{A}$, $\mathcal{B}$)-controlled invariant contained in a given subspace $\mathcal{X}$, which is shortly denoted by MCI($\mathbf{A}$, $\mathcal{B}$, $\mathcal{X}$), is provided by the following sequence of subspaces, which converges at most in n-1 steps.

$$\mathscr{L}_0 = \mathscr{X}$$

$$\mathscr{L}_i = \mathscr{X} \cap \mathbf{A}^{-1*}(\mathscr{L}_{i-1} + \mathscr{B}) \qquad (i=1,2,..) \ .$$

Algorithm 4: The minimum $(\mathbf{A},\mathscr{B})$-conditioned invariant containing a given subspace $\mathscr{X}$, which is shortly denoted by mci$(\mathbf{A},\mathscr{B},\mathscr{X})$, is provided by the following sequence of subspaces, which converges at most in n-1 steps.

$$\mathscr{L}_0 = \mathscr{X}$$

$$\mathscr{L}_i = \mathscr{X} + \mathbf{A}(\mathscr{L}_{i-1} \cap \mathscr{B}) \qquad (i=1,2,..) \ .$$

Remark: that Algorithms 1 and 2 are special cases of Algorithms 3 and 4, corresponding to $\mathscr{B} = \mathbf{0}$ and $\mathscr{B} = R^n$ respectively.

B.Constrained controllability.

Consider the input-state differential equations

$$\dot{\mathbf{x}}(t) = \mathbf{A}\mathbf{x}(t) + \mathbf{B}\mathbf{u}(t) \quad , \qquad (23)$$

where $\mathbf{x} \in R^n$, $\mathbf{u} \in R^m$. The most important property which links the control of the system (22) to controlled invariants is the following.

Property 1: A subspace $\mathscr{I} \subseteq R^n$ is a $(\mathbf{A},\mathscr{R}(\mathbf{B}))$-controlled invariant if, and only if, for any initial state $\mathbf{x}_0 \in \mathscr{I}$, there exists at least a control action $\mathbf{u}(t)$, $t \geqslant 0$, such that the corresponding state trajectory $\mathbf{x}(t)$, $t \geqslant 0$, of the system (22) ranges over $\mathscr{I}$.

Remark that, according to Property 1, given a subspace $\mathscr{X} \subseteq R^n$, the subset $\mathscr{X}_1 \subseteq \mathscr{X}$ of the initial states from which trajectories of the system (22) completely ranging on $\mathscr{X}$ can be obtained by proper control actions is the subspace MCI$(\mathbf{A}, \mathscr{R}(\mathbf{B}),\mathscr{X})$. Of course, every trajectory ranging over $\mathscr{X}$ ranges over $\mathscr{X}_1$. Every couple of states belonging to $\mathscr{X}_1$ are not necessarily connectable by a trajectory ranging over $\mathscr{X}_1$ if they do not belong to a smaller subspace $\mathscr{X}_2 \subseteq \mathscr{X}_1$, the set of states reachable from the origin by trajectories ranging over $\mathscr{X}_1$, which will be called the maximum reachable subspace on $\mathscr{X}$. An algorithm for the computation of this subspace is easily derived by using the following further properties of controlled invariants.

Property 2: For any $(\mathbf{A}, \mathscr{R}(\mathbf{B}))$-controlled invariant $\mathscr{I}$ there exists at least one matrix $\mathbf{H}$ such that $(\mathbf{A}+\mathbf{BH})\mathscr{I} \subseteq \mathscr{I}$.

Property 3: The maximum reachable subspace on a $(\mathbf{A}, \mathscr{R}(\mathbf{B}))$-controlled invariant $\mathscr{I}$ is the subspace of the reachable states of the system

$$\dot{\mathbf{x}}(t) = (\mathbf{A}+\mathbf{BH})\ \mathbf{x}(t) + \mathbf{G}\mathbf{v}(t) \quad , \qquad (24)$$

where $\mathbf{G}$ is any matrix such that $\mathscr{R}(\mathbf{G}) = \mathscr{I} \cap \mathscr{R}(\mathbf{B})$ and $\mathbf{H}$ is any matrix such that $(\mathbf{A}+\mathbf{BH})\mathscr{I} \subseteq \mathscr{I}$.

Algorithm 5: The maximum reachable subspace contained in a general subspace $\mathscr{X} \subseteq R^n$, which is shortly denoted by MRS $(\mathbf{A}, \mathscr{R}(\mathbf{B}), \mathscr{X})$, can be computed as $mi(\mathbf{A}+\mathbf{BH}, \mathscr{I} \cap \mathscr{R}(\mathbf{B}))$, being $\mathscr{I} = MCI(\mathbf{A}, \mathscr{R}(\mathbf{B}), \mathscr{X})$ and $\mathbf{H}$ any matrix such that $(\mathbf{A}+\mathbf{BH})\mathscr{I} \subseteq \mathscr{I}$.

Property 4: Given a $(\mathbf{A}, \mathscr{R}(\mathbf{B}))$-controlled invariant $\mathscr{I}$ completely reachable, i.e. such that $MRS(\mathbf{A}, \mathscr{R}(\mathbf{B}), \mathscr{I}) = \mathscr{I}$ there exists at least one matrix $\mathbf{H}$ such that $\mathscr{I}$ is a $(\mathbf{A}+\mathbf{BH})$-invariant and has arbitrary associated eigenvalues.

Property 5: Given a $(\mathbf{A}, \mathscr{R}(\mathbf{B}))$-controlled invariant $\mathscr{I}$ and any linear map $\mathbf{P}$ from R^n such that $\mathscr{N}(\mathbf{P}) \cap \mathscr{I} = \mathbf{o}$, there exists at least one matrix $\mathbf{H'}$ having proper dimensions such that $(\mathbf{A}+\mathbf{BH'P})\mathscr{I} \subseteq \mathscr{I}$. Furthermore arbitrary eigenvalues can be associated to the $(\mathbf{A}+\mathbf{BH'P})$-invariant $\mathscr{I}$ if it is completely reachable.

REFERENCES

1. G.Basile,R.Laschi and G.Marro, "Invarianza controllata e non interazione nello spazio degli stati", *L'Elettrotecnica*, vol.56, n.1, pp.14-18, January 1969.
2. G.Basile and G.Marro, "Controlled and conditioned invariant subspaces in linear system theory", *Journal of Optimization Theory and Applications*, vol.6, n.5, pp.410-415, May 1969.
3. M.W.Wonham and A.S.Morse, "Decoupling and pole assigment in multivariable systems: a geometric approach', *SIAM Journal on Control*, vol.8, n.1, pp.1-18, January 1970.

4. G.Basile and G.Marro, "Relazione fra la stabilità e la controllabilità dei sistemi lineari su alcuni sottospazi caratteristici", *L'Elettrotecnica*, vol.57, n.2,pp.61-66, February 1970.
5. G.Basile and G.Marro, "A state space approach to non-interacting controls", *Ricerche di Automatica*, vol.1,n.1, pp.68-77, September 1970.
6. M.W.Wonham, "On pole assigment in multi-input controllable linear systems", *IEEE Transactions on Automatic Control*, vol.AC-12, n.5, pp.660-665, December 1967.
7. D.G.Luenberger, "Observing the state of a linear system" *IEEE Transactions on Military Electronics*, vol. MIL - 8 n.2, pp.74-80, April 1964.
8. D.G.Luenberger, "Observers for multivariable systems", *IEEE Transactions on Automatic Control*, vol.AC-11, n.2, pp.190-197, April 1966.
9. D.G.Luenberger, "An introduction to observers", *IEEE Transactions on Automatic Control*, vol.AC-16, n.6, pp. 596-602, December 1971.
10. G.Basile and G.Marro, "L'invarianza rispetto ai disturbi studiata nello spazio degli stati", presented at the 70th annual meeting of the Italian Electrotechnical and Electronic Association, Rimini, Italy, September 1969, Paper n.1.4.01.
11. G.Basile and G.Marro, "A new characterization of some structural properties of linear systems: unknown-input observability, invertibility and functional controllability" - to be published.
12. G.Basile and G.Marro, "On the synthesis of unknown-input observers and inverse systems by recursive algorithms"- to be published.

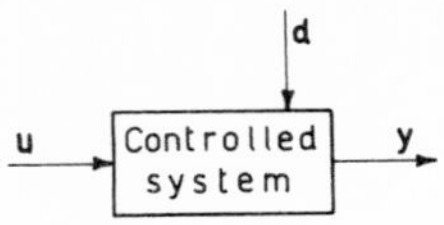

a)

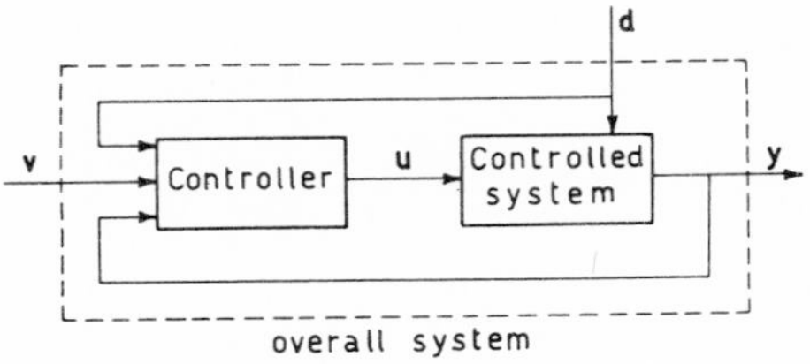

b)

Fig. 1 - The considered control system

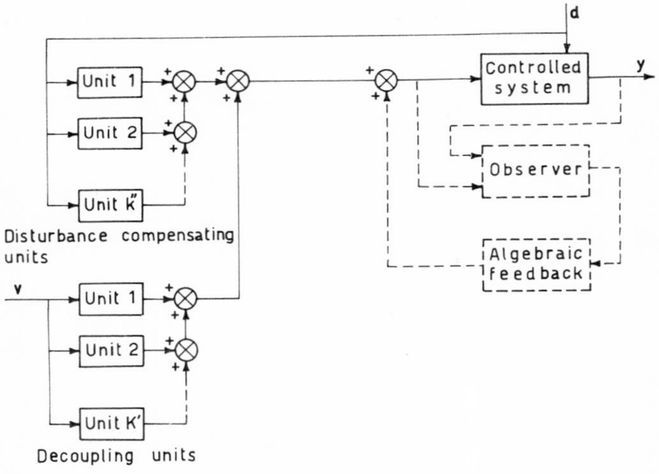

Fig. 2 - The structure of the controller achieving decoupling and disturbance insensitivity.

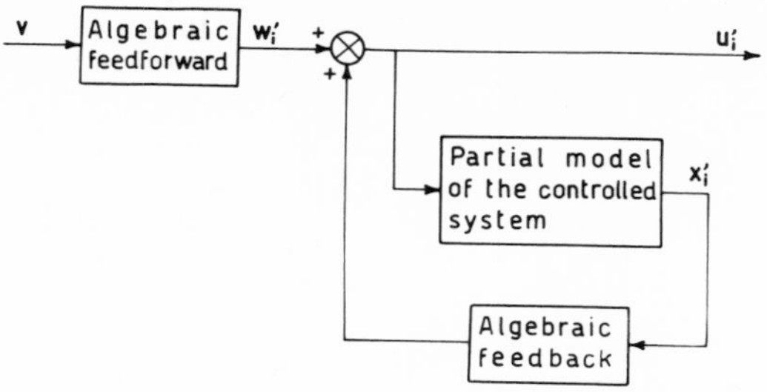

Fig. 3 - A decoupling unit.

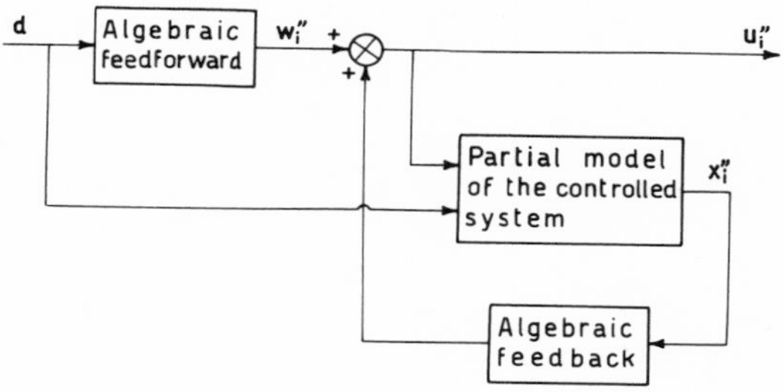

Fig. 4 - A disturbance compensating unit.

THE REALIZATION OF NEAR-OPTIMAL FEEDBACK CONTROLLERS

Harold K. Knudsen

Department of Electrical Engineering
and Computer Science
The University of New Mexico
Albuquerque, New Mexico

1. Introduction

The engineering problem which must be solved in the design of (near-) optimal feedback controllers is not: Specify the optimal controller, but rather is: Specify, from a class of controllers of given complexity, that controller which gives the best performance. This is a multi-faceted problem that includes the problems of approximation of optimal control laws and controller realization. The class of systems considered are described by ordinary differential equations with bounded inputs and integral performance indices. The function of the controller is to steer the initial state to a fixed target. It is assumed that optimal feedback control laws exist and that these control laws can be approximated by a multilevel feedback control law.

The two main facets of the design problem, the approximation of optimal control laws through the development of near-optimal control laws, and controller realization, are developed separately.

Near-optimal control laws, which form a central conceptual basis for this paper, are control laws which guarantee a lower bound on the level of performance of the controlled system. More specifically, near-optimal control laws are defined as follows. Let $u = \psi(\underline{x})$ be a feedback control law defined on a region Q in state space. Let $J_\psi(\underline{x})$ be the cost of transfer from $\underline{x}$ to the target using the control law $u = \psi(\underline{x})$, and let $V(\underline{x})$ be the optimal cost of transfer $\underline{x}$ to the target. The control law $u = \psi(\underline{x})$ is said to be near-optimal with level of performance p in Q, if for all $\underline{x}\varepsilon Q$, $pJ_{\psi}(\underline{x}) \leq V(\underline{x})$. p is a real number which satisfies $0 < p \leq 1$. As p approaches one, the near-optimal policies approach an optimal policy. Two approaches for obtaining near-optimal feedback control laws are stated. The first, an algorithm based on policy improvement, is of general application while the second, which is of use in the case of bang-bang control, is based on a sufficient condition for near-optimality.

The realization of multilevel feedback controllers is considered in section 4 where properties of the given class of controllers are derived. The minimal realization of the controller logic is discussed briefly for the case of bang-bang control laws which dichotomize the state space.

Section 5 combines the material on near-optimal control and controller realization by presenting simulation results on a controller design.

2. Problem Statement

Near-optimal feedback controllers are to be realized for the following regulator type problem:

System dynamics. The dynamics of the system to be controlled are assumed to be modeled by the differential equation

$$\dot{\underline{x}} = \underline{f}(\underline{x},u), \ \underline{x}\varepsilon E^n, \ |u| \leq 1 \tag{2.1}$$

where the state space E^n is an n-dimensional Euclidean space and the control u is scalar.

Controller function. The purpose of the feedback controller is to generate a control law $u = \psi(\underline{x})$ which transfers any initial state $\underline{x}$ belonging to a specified Q† called the initial state set to a target S in finite time. The existence of such a control policy will be assumed.

System Trajectories. It is assumed that the control law $u = \psi(\underline{x})$ is chosen such that equation (2.1) has a unique solution

$$\underline{x}(t) = \underline{\Gamma}_{\psi}(t;\underline{x}_o) \tag{2.2}$$

† $Q\varepsilon E^n - S$. The control law $\psi(\underline{x})$ must, in general, be specified for states not contained in Q.

for all initial states in Q and $t \geq 0$. The curve $\Gamma(\cdot;\underline{x}_o)$ will be called a trajectory.

Index of performance. The cost of transfer of an initial state $\underline{x}_o \varepsilon Q$ to S using the control law $u = \psi(\underline{x})$ is taken to be

$$J_\psi(\underline{x}_o) = \int_o^{t_1} L[\underline{x},\psi(\underline{x})]dt \tag{2.3}$$

where $\underline{x}(t) = \underline{\Gamma}_\psi(t;\underline{x}_o)$ and t_1 is the first time at which the trajectory emanating from $\underline{x}_o$ intercepts the target S.††

Further assumptions on system dynamics and the index of performance. In order to restrict the problem of controller realization, it will be assumed that $L[\underline{x},\psi(\underline{x})]$ can be chosen such that the optimal control law is (or can be adequately approximated by) a multilevel control policy. Furthermore, it is assumed that an optimal feedback control law $\psi^*(\underline{x})$ exists.

3. Near-Optimal Feedback Control Laws

The definition of near-optimal feedback control laws, their interpretation in terms of cost controllable sets, and methods for determining these control laws are presented in this section.

Definition of near-optimal control. A feedback control law $u = \psi(\underline{x})$ will be called near-optimal with performance level p on Q if $pJ_\psi(\underline{x}) \leq V(\underline{x})$ $\forall$ $\underline{x}\varepsilon Q$ where the constant p satisfies $0 < p \leq 1$ and $V(\underline{x}) = J_{\psi^*}(\underline{x})$ where $\psi^*(\underline{x})$ is the optimal feedback control law. Within this definition, as p approaches 1, the near-optimal policy approaches optimality, while $p \to 0$ implies no more than a finite cost of transfer.

Cost-controllable sets. A geometric interpretation of near-optimality as well as an alternate definition can be obtained in terms of the cost-controllable set $A_\psi(c)$ defined by

$$A_\psi(c) = \{\underline{x}\varepsilon Q: \; J_\psi(\underline{x}) \leq c\} \quad .$$

††$L[\underline{x},\psi(\underline{x})]$ is assumed non-negative.

If L, the integrand in the index of performance, is non-negative, $A_\psi(c_1) \subset A_\psi(c_2)$ for all c_1, c_2 satisfying $0 \leq c_1 \leq c_2$.

Near optimality is expressed in terms of cost-controllable sets by introducing the optimal cost-controllable set

$$A^*(c) = \{\underline{x} \varepsilon Q: \quad V(\underline{x}) \leq c\} \quad .$$

If $A^*(pc) \subset A_\psi(c)$ $\forall$ $c \varepsilon [0, c_1]$, then ψ is near-optimal with performance level p on $A_\psi(c_1)$.

Proof. Let $\underline{x_1}$ be an arbitrary point in $A_\psi(c_1)$. Put $c = V(\underline{x_1})/p$. By hypothesis, $\underline{x_1} \varepsilon A^*(pc) \Rightarrow \underline{x_1} \varepsilon A_\psi(c) \Rightarrow J_\psi(\underline{x_1}) \leq c$ or $pJ_\psi(\underline{x_1}) \leq V(\underline{x_1})$.

Methods for determining near-optimal feedback control laws. Two methods for determining near-optimal feedback control laws are outlined below. The first, policy improvement on the cost controllable sets, is an application of dynamic programming [1], and is of general application. The second method, which makes use of sufficient conditions of near-optimal control, is applied to the near-time-optimal control of linear systems where interesting results are obtained.

The policy improvement approach is based on the cost controllable sets $A_\psi(c)$. The goal of policy improvement is to generate from a given near-optimal policy $\psi_o(\underline{x})$, a sequence of near-optimal control policies $\psi_o(\underline{x})$, $\psi_1(\underline{x})$,..., with non-decreasing levels of performance p_0, p_1, p_2,..., respectively, which converges to an optimal policy. To achieve this goal, policy improvement will be defined as follows: Let $\psi_i(\underline{x})$ be a given policy. $\psi_{i+1}(\underline{x})$ will be said to be an improvement on the policy $\psi_i(\underline{x})$ on the set $A_\psi(c)$ if $A_{\psi_i}(c) \subset A_{\psi_{i+1}}(c)$.

If an improvement is made at each step, the sequence $A_{\psi_o}(c) \subset A_{\psi_1}(c) \subset \ldots$, converges to $A_{\psi^*}(c)$, and the sequence of feedback control policies $\psi_o(\underline{x})$, $\psi_1(\underline{x})$, ..., converges to an optimal policy $\psi^*(\underline{x})$ on the set $A_{\psi^*}(c)$. It follows from the definition of near-optimal policies that the sequence p_0, p_1, p_2, ..., is non-decreasing and converges to the value 1.†

†Even if an optimal policy doesn't exist, this method still generates a sequence of improving control laws. In any case, the process is terminated when satisfactory performance is achieved.

Policy improvements made within the constraints of a given controller structure will not, in general, yield the optimal policy but will, at least in theory, indicate if the presumed controller structure is adequate to realize a feedback control law with the desired level of performance. In practice, application of policy improvement algorithms is limited by computational feasibility. For the near-optimal control problem, computational feasibility is governed by:

a) the complexity of the relation between controller parameters and near-optimal feedback control policies, and
b) the quality (or existence) of a cost approximation theory.

Unfortunately, for multilevel control policies, only rudimentary knowledge exists in either of these areas. Thus, policy improvement is now of little value in the solution of this class of problem.

The second method for determining near-optimal feedback control laws, which is based on a sufficient condition for near-optimality, requires that the optimal cost $V(\underline{s})$ can be evaluated at any point in Q. This sufficient condition is developed below. If the control law $\psi(\underline{x})$ is near-optimal with performance level p on Q, then from equation (2.3)

$$p \int_0^{t_1} L[\underline{x}, \psi(\underline{x})]\,dt \leq V(\underline{x}_o) \tag{3.1}$$

where $\underline{x}(t) = \underline{\Gamma}_\psi(t;\underline{x}_o)$.

Let $W^*(t;\underline{x}_o,\psi) \triangleq V[\underline{\Gamma}_\psi(t;\underline{x}_o)]$. (3.2)

Since $W^*(t_1;\underline{x}_o,\psi) = 0$, condition (3.1) can be written using equation (3.2) as

$$p \int_0^{t_1} L[\underline{x}, \psi(\underline{x})dt \leq W^*(0;\underline{x}_o,\psi) - W^*(t_1;\underline{x}_o,\psi) \tag{3.3}$$

If W^* exists, condition (3.3) can be written as

$$p\int_{0}^{t_1} L[\underline{x},\psi(\underline{x})]dt \leq -\int_{0}^{t_1} \dot{W}^*(t;\underline{x}_o,\psi)dt \quad . \tag{3.4}$$

A sufficient condition for (3.4) to hold is

$$pL[\underline{x},\psi(\underline{x})] \leq -\dot{W}^*(0;\underline{x},\psi) \tag{3.5}$$

for all $\underline{x}$ belonging to the trajectory $\Gamma_{\psi}(t,\underline{x}_o)$, $t\varepsilon[0,t_1]$. Thus, if a $\psi(\underline{x})$ can be found such that (3.5) is satisfied for all $\underline{x}$ in the trajectories emanating from Q to S, $\psi(\underline{x})$ is at least of performance level p on Q.

Andreas [2] has applied this condition to problems in which the optimal control is bang-bang. This leads to an S_p set characterization of the near-optimal control law which is described below.

If the optimal control law $\psi^*(\underline{x})$ is bang-bang on E^n-S, then the following sets, denoted jointly as S_p sets, can be defined:

$$S_p^+ \triangleq \{\underline{x}\varepsilon E^n-S:\ \psi^*(\underline{x}) = +1, \exists\, a\,|u| \leq 1 \text{ such that } pL(\underline{x},u) > -\dot{W}^*(0;\underline{x},u)\}$$

$$S_p^- \triangleq \{\underline{x}\varepsilon E^n-S:\ \psi^*(\underline{x}) = -1, \exists\, a\,|u| \leq 1 \text{ such that } pL(\underline{x},u) > -\dot{W}^*(0;\underline{x},u)\}$$

A near-optimal control law with a performance level not less than p can be written directly in terms of the S_p sets as

$$\psi(\underline{x}) = \begin{cases} +1, & \underline{x}\varepsilon\ S_p^+ \\ u,\ |u| \leq 1, & \underline{x}\varepsilon\ E^n-(S_p^+ \cup S_p^-) \\ -1, & \underline{x}\varepsilon\ S_p^- \end{cases}$$

The S_p sets for the near-time-optimal control of the linear constant controllable system $\dot{\underline{x}} = \underline{A}\ \underline{x} + \underline{b}\ u$, $|u| \leq 1$ where $\underline{A}$ is negative definite and the target S is a (hyper-) sphere centered at the origin have been shown to have the following properties [2]:

P1: S_p^+ and S_p^- are compact sets.

P2: $S_p^+ \cap S_p^- = \emptyset$ for $0 < p < 1$.

P3: S_p^+ and S_p^- are strictly nested as p varies. That is, if $0 < P_2 < P_1 < 1$, then $\partial S^+_{p1} \cap S^+_{p2} = \emptyset$ and similarly, $\partial S^-_{p1} \cap S^-_{p2} = \emptyset$.

P4: For $0 < P < 1$, a surface Σ exists which strictly separates S^+_p from S^-_p.

The above properties guarantee the existence of a separating surface $\sum$ which can be used as a switching surface in the realization of a near-optimal feedback controller (as in Fig. 3.1). Property P3 implies that, as p is made smaller (which corresponds to a worse bound on the sub-optimal controls), the S_p sets become smaller and hence easier to separate. Thus, varying the parameter p allows a trade-off between controller complexity and performance.

<u>S_p Set description for the time optimal problem.</u> The S_p sets are approximated by computing a grid of points contained in the sets. Fig. 3.2, which is taken from Andreas [2] shows the S_p sets for the time-optimal control of the system

$$\underline{\dot{x}} = \begin{bmatrix} -0.1 & 1 \\ -1 & -0.1 \end{bmatrix} \underline{x} + \begin{bmatrix} 0 \\ 1 \end{bmatrix} u, \quad \underline{u} \leq 1, \ S = \{\underline{x}: \|\underline{x}\| = 0.1\}$$

for p = 0.8 and 0.4. The figure demonstrates the nesting property of the S_p sets. These sets were computed using reverse time flooding. The performance of controllers based on S_p set operation and using the controller structure given in the next section is described in section 5.

4. Controller Structure and Realization

<u>Controller structure.</u> Piecewise constant control laws of the form $\psi(\underline{x}) = u_i$, $\underline{x} \varepsilon C_i$, $i = 1, \ldots, r$ where the <u>control sets</u> are subsets of state space which satisfy $\bigcup_{i=1}^{r} C_i = E^n$, and $C_i \cap C_j = \emptyset \ \forall \ i, j; \ i \neq j$ can be realized using the feedback controller structure shown in Fig. 4.1. The controller, as shown in Fig. 4.1, consists of a mapping $\underline{\Phi}(\underline{x})$, which maps state space into a (higher dimensional) space Y. $\underline{\Phi}$ is assumed a one-to-one mapping from E^n into Y. The point $\underline{y} = \underline{\Phi}(\underline{x})$ is then operated on by the threshold logic units (TLUs) to form a Boolean code $\underline{z}$. The TLUs act to partition the space Y by defining the half-spaces

$H_i^+ \triangleq \{\underline{y} : \underline{a}_i^T \underline{y} \geq b_i\}$ and

$$H_i^- \triangleq \{\underline{y} : \underline{a}_i^T \underline{y} < b_i\} \quad , i = 1, \ldots, \ell.$$

The TLU outputs $z_1, \ldots, z_\ell$ are

$$z_i = \begin{cases} 1, \underline{y} \; \varepsilon H_i^+ \\ 0, \; \underline{y} \; \varepsilon H_i^- \end{cases} \qquad i = 1, \ldots, \ell.$$

the LOGIC block assigns a control u_i to each Boolean code $\underline{z} = [z_i, z_2, \ldots, z_\ell]$.

The control sets, $C_1, C_2, \ldots, C_r$, over which the corresponding constant controls $u_k, u_2, \ldots, u_r$ are generated, are realized as follows. Let the non-empty intersections of the half-spaces defined by the TLUs be denoted by $F_i \triangleq H_1{}^{\alpha_{i1}} \cap \ldots \cap H_\ell{}^{\alpha_{i\ell}}$, $i = 1, \ldots, r$ where $\alpha_{ij} \; \varepsilon\{+,-\}$. Then the control set C_i can be defined as

$$C_i = \underline{\Phi}^{-1}(F_i \cap \underline{\Phi}(E^n)), \; i = 1, \ldots, r. \tag{4.1}$$

The control sets, as defined by equation (4.1) satisfy the conditions that their union covers E^n and that the intersection of any distinct pair is empty. It may happen, however, that a given C_i is empty or has more than one component. From equation (4.1) and the definition of F_i, it is clear that an alternate expression for the control sets can be made in terms of the preimages of $H_j^{\alpha_j} \cap B$. That is, if

$$P_j^\alpha \triangleq \underline{\Phi}^{-1}(H_j^\alpha \cap B), \; \alpha\varepsilon\{+,-\} \quad \text{then}$$

$$C_i = \bigcap_{j=1}^{\ell} P_j{}^{\alpha_{ij}} \quad \text{if} \quad F_i = \bigcap_{j=1}^{\ell} H_j{}^{\alpha_{ij}} \quad .$$

Note that, from the above definition, P^+ and P^- can be expressed simply as

$$P_j^+ = \{\underline{x} \; \varepsilon E^n : \underline{a}_j^T \; \underline{\Phi}(\underline{x}) \geq b_j\}$$

and

$$P_j^- = \{\underline{x} \; \varepsilon E^n : \underline{a}_j^T \; \underline{\Phi}(\underline{x}) < b_j\} \quad .$$

In what follows, it will be useful to define by

$$S_j = \{\underline{x} \in E^n : \underline{a}_j^T \underline{\Phi} (\underline{x}) = b_j\}$$

the (hyper-) surface which separates P_j^+ from P_j^-.

The controller structure described above acts (between the state input $\underline{x}$ and the TLU output $\underline{z}$) to assign to each state $\underline{x}$ in E^n a Boolean code $\underline{z}$.

It is convenient to view the controller operation as a mapping $\underline{z} = \underline{M}(\underline{x})$ by which each control set C_i is mapped into a vertex of a unit hypercube in the TLU output space E^ℓ [3,4,5]. Fig. 4.2 illustrates this concept by showing the control sets (and respective Boolean codes) defined by three surfaces S_1, S_2, and S_3 in state space (E^2) in Fig. 4.2(a) and the images of the control sets on the unit hypercube in the TLU output space (E^3) in Fig. 4.2(b).

LOGIC realization. The block (in Fig. 4.1) which assigns to the unique Boolean code $\underline{z}_1$ (associated with the control set C_i) is certainly realizable, but apparently no work has been done on minimal realizations for the multi-level control problem. However, if a bang-bang control is to be realized,† then $\underline{M}$ has added properties which are enumerated below and which make the minimal realization of the LOGIC block more feasible and which will be called the switching surface realization problem.

The switching function realization problem. Let $C^+ = \{\underline{x} : \underline{x} \in C_i, u_i = +1, i = 1, \ldots, r\}$ denote the union of all positive control sets and $C^- = \{\underline{x} : \underline{x} \in C_i, u_i = -1, i = 1, \ldots, r\}$ denote the union of all negative control sets. If C^+ is connected and C^- is connected, C^+ and C^- can be separated by a switching surface. If conditions (a) and (b) (which follow) are satisfied, then both C^+ and C^- are connected. These conditions are:

a) P_j^+ is connected and P_j^- is connected $j = 1, \ldots, \ell$.
b) The images of the positive (negative) control sets in the TLU output space form, on adding edges of the hypercube between adjacent†† vertices of the

---

†The bang-bang control policy considered here is one in which the sets $\{\underline{x} \in E^n - S : \psi(\underline{x}) = 1\}$ and $\{\underline{x} \in E^n - S : c(\underline{x}) = -1\}$ dichotomize the $E^n - S$ into two disjoint connected sets.

††Vertices of distance 1 are adjacent.

same category (both vertices, corresponding to control sets in C^+ (or C^-)), connected subgraphs. One subgraph corresponds to the control sets in C^+, the other control sets in C^-.

With these added restrictions on $\underline{M}$, and the additional requirement that $\underline{\Phi} = \underline{I}$, some research, primarily in pattern recognition, has been done in minimal realization [4,6,7].

The author has obtained results on the minimal TLU realization based on a linear programming approach. The LP approach is computationally feasible because of the sparseness of the Boolean function associated with the switching surface.

5. Controller Design and Simulation Results

Experimental near-time-optimal feedback controllers have been realized using a multi-surface separation of points in the S_p sets [8]. Simulation indicates that the controllers give excellent performance for second order systems and reasonable performance for third order linear systems. The complexity of the controllers obtained (number of surfaces required for separation) is, with the algorithm employed, too great for practical use in the case of third order systems.

Example. A controller for near-time-optimal control of the system

$$\dot{\underline{x}} = \begin{bmatrix} -0.1 & 1.0 & 0 \\ -1.0 & -0.1 & 0 \\ 0 & 0 & -0.01 \end{bmatrix} \underline{x} + \begin{bmatrix} 0 \\ 1 \\ 1 \end{bmatrix} u, \quad |u| \leq 1$$

with target radius 0.1 on the set $Q = \{\underline{x} \in E^3 - S : \|\underline{x}\| \leq 1\}$, was realized by separating 772 points in the sets $S^+_{0.8}$ and $S^-_{0.8}$. Separation was performed using a heuristic algorithm which acted to separate, using a quadric surface, the points in the S_p sets contained in a given cube in state space. If separation proved impossible, the cell was subdivided through bisection and quadric separation was again attempted. The resulting controller required 13 quadric surfaces plus logic to determine cell location. This might seem extreme but, as shown in Fig. 5.1 (from Andreas), the S_p sets for this system are rather complex. The simulation responses $\bar{t}$ from the initial states indicated are listed on the following page and compared to the optimal response times t^*.

Simulated Response Times

x_1	x_2	x_3	$\bar{t}$	t^*	$\bar{t}/t^*$
0.5	-0.5	-0.5	2.025	0.9941	2.04
0.25	-0.75	-0.75	0.876	0.6929	1.26
0.75	-0.25	-0.75	1.186	1.380	1.31
0.25	-0.25	-0.25	1.022	0.581	1.76
0.125	-0.125	-0.125	0.854	0.223	3.83
0.125	-0.375	-0.375	0.728	0.323	2.25
0.375	-0.125	-0.375	1.552	0.959	1.49
0.187	-0.062	-0.187	1.069	0.567	1.88

The design goal was to make the response time $\bar{t} \leq 1.25*$. The deviations from this goal are apparently the result of too crude of an approximation to the S_p sets. The approximation can be approved by using a larger number of points to describe the S_p sets.

Studies on computational feasibility indicate that near-time-optimal controller realization for systems through fourth order is practical with the above method [8]. Both computer time and storage become excessive for most fifth and higher order systems.

6. Conclusions

A procedure has been outlined for the design of near-optimal feedback controllers. Near-optimal control policies are defined and characterized in terms of cost-controllable sets. It is shown that near-optimal control policies can be obtained through policy improvement. In the case of bang-bang control, sufficient conditions for near-optimal control are specified in terms of S_p sets. Separation of the S_p sets with a switching surface leads to near optimal feedback controllers. A near-time-optimal feedback controller was realized through approximate separation of S_p sets, and the resulting controlled system's performance was tested. A general structure for the realization of piecewise constant feedback controllers and elementary properties of the realization have also been described.

The most important area for future research in the field of near-optimal control is the development of an approximation theory for $J_\psi(\underline{x})$, the cost of transferring the state $\underline{x}$ to the target with feedback control $u = \psi(\underline{x})$. Such a theory would make it feasible to use policy improvement to generate near-optimal feedback control policies. It would also allow, by specifying restrictions on the control policy, a more direct relation between controller parameters and the near-optimal control policy.

Acknowledgments

The author wishes to acknowledge the work of his students, Chon Ho Hyon and Chaw-Kwei Hung, especially Mr. Hung's assistance in writing the computer simulation programs. This paper would not have been possible without the work of Dr. Ronald Andreas on S_p sets and Dr. Anthony Veneruso on controller realization. The author gratefully acknowledges their contribution.

References

1. R. Bellman, Dynamic Programming, (1957).
2. Andreas, R. D., An Approach for Suboptimal Feedback Control, Ph.D. dissertation, The University of New Mexico, January 1970.
3. N. J. Nilsson, Learning Machines, (1965).
4. A. F. Veneruso, Realization of Near-Time-Optimal Control, Ph.D. dissertation, The University of New Mexico, January 1971.
5. H. K. Knudsen, "Research on the Design of Near-Optimal Controllers," The University of New Mexico, Technical Report EE-188(70)SAN-183, June 1971.
6. A. G. Arkader, and E. M. Braverman, Computers and Pattern Recognition, (1967).
7. F. W. Smith, IEEE Trans. on Comp., 17, 367-372, (1968).
8. H. K. Knudsen, "The Realization of Near-Time-Optimal Feedback Controllers," The University of New Mexico, Technical Report EE-194(72)SAN-217, January 1972.

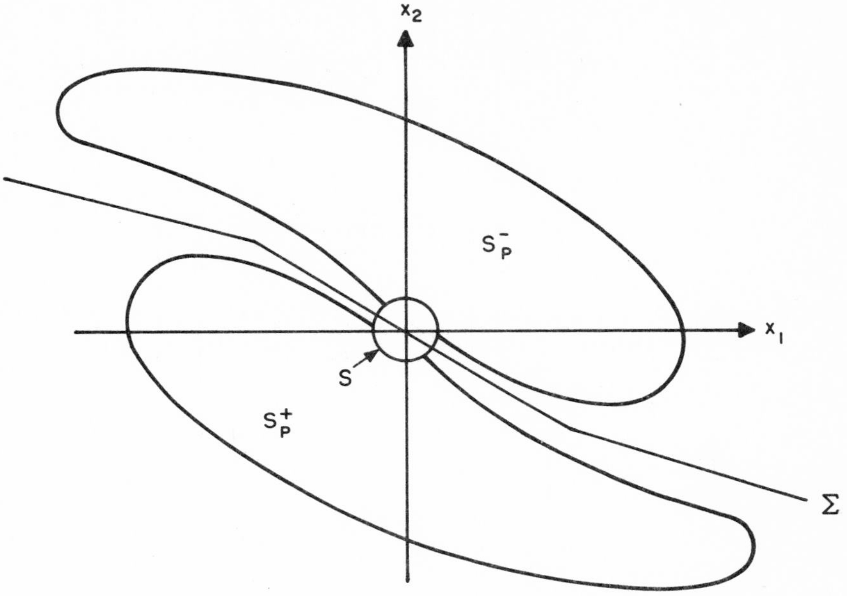

Fig. 3.1. The Switching Surface Realization of a Near-Optimal Controller

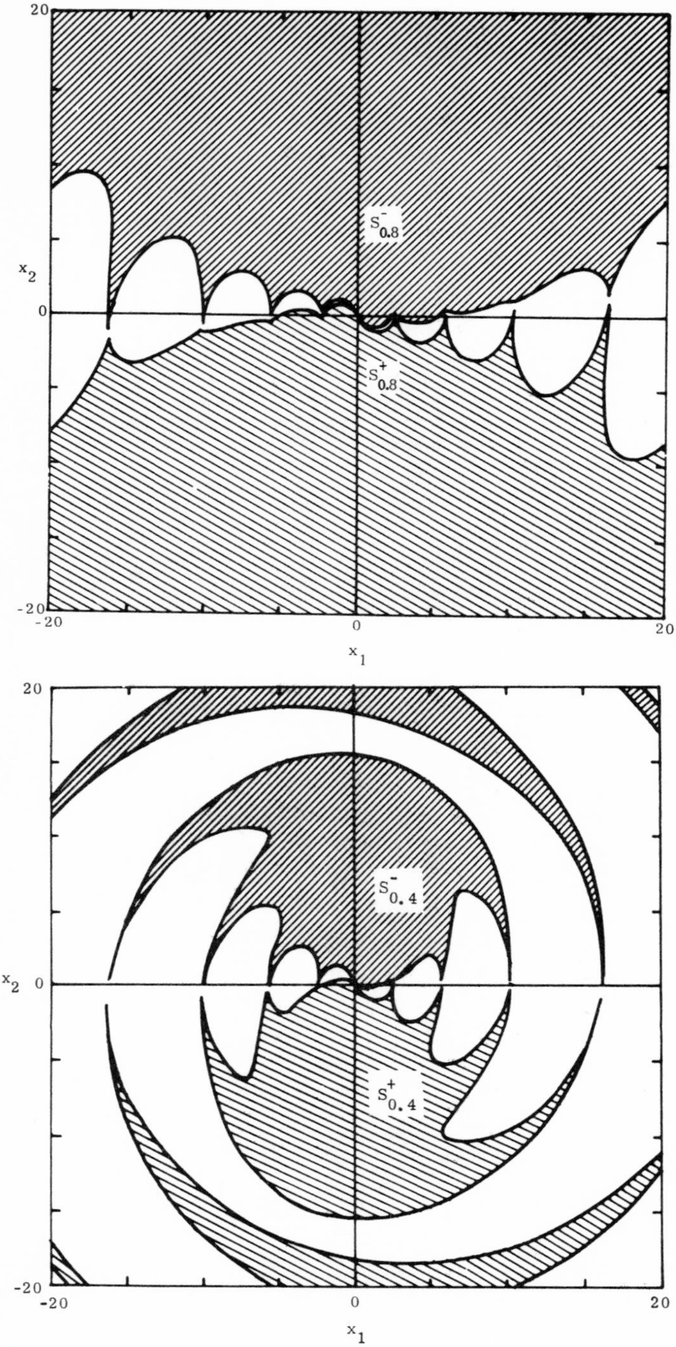

Fig. 3.2. S_p Sets for a Damped Harmonic Oscillator

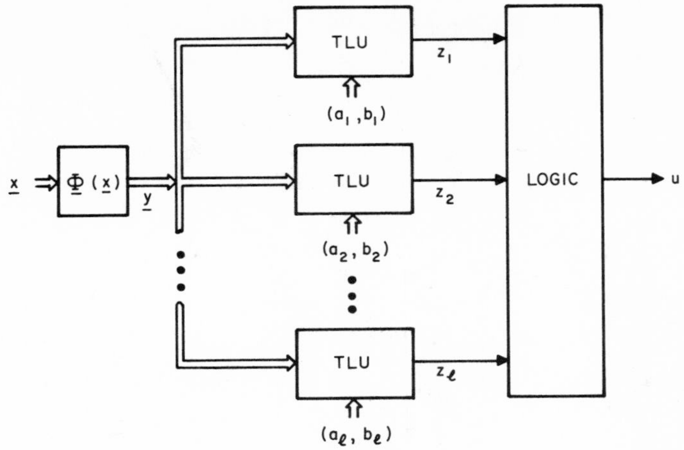

Fig. 4.1. Feedback Controller Structure

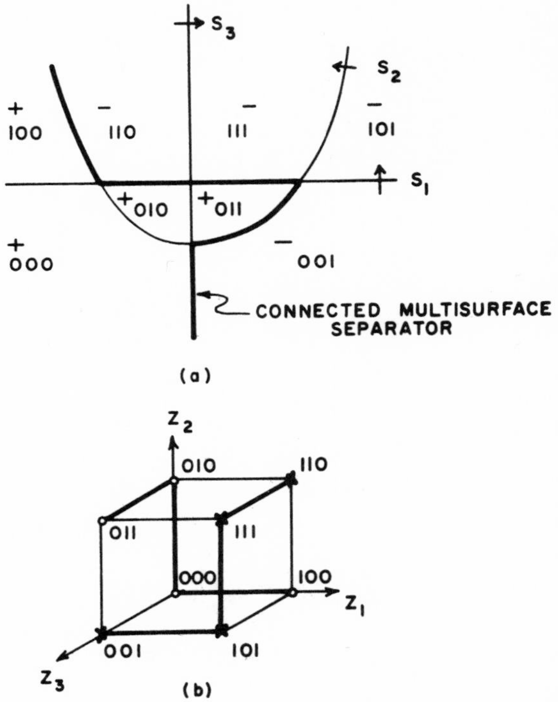

Fig. 4.2. A Three-Surface Separator in E^2 and Its Image in the TLU Output Space

Fig. 5.1. $S_{0.4}$ Set Surfaces for the Third Order System

THE BASE MODEL CONCEPT

Bernard P. Zeigler*

Logic of Computers Group
Department of
Computer and Communication Sciences
The University of Michigan
Ann Arbor, Michigan 48104

Abstract

In this paper, the concept of *base model* is introduced to clarify the large gap between the structural level at which true understanding of a system is believed to reside and the level at which computer simulations are possible. As an applications example, the concept of structure morphism is used to derive simulatable brain models from assumed base models at the neuron net level.

Introduction

In many areas of modelling and simulation there is a large gap between the (micro) level of system structure at which true understanding is believed to reside and the (macro) level at which computer simulation is feasible. Examples of such areas are the simulation of living bacterial cells and the simulation of ecosystems. Biologists usually consider molecular coordinates to be appropriate for a complete description of the cell but feasible computer simulations (of global cell behavior, not just individual pathways) must necessarily relate to highly aggregated pools of biochemicals (Zeigler and Weinberg, 1970). Similarly, ecologists conceive of significant ecological

---

*This work was supported by the National Institutes of Health, Grant No. GM-12236 and the National Science Foundation, Grant No. GJ-29989X.

interactions as occurring among the individual species sharing a habitat but computer simulations necessarily deal with lumped compartments of organism types. Again, the reference here is to simulations of global ecosystem behavior, (Patten, 1972; see also his paper in the present volume).

It is the contention of this paper that the gap, between what the modeller wants to know and what he can actually deal with, can be better understood within a formal theory of modelling and simulation. As in previous formulations (Zeigler, 1971a,b; 1972a,b), a simulation involves three elements: a real system, a model and a computer. However this picture is now to be refined (as in Figure 1) by introducing the notion of *base model* as an essential component. The *real system* is assumed to be knowable only through behavioral observations and hence is represented by a set R of input-output segment pairs. R is taken to be complete in the sense that all possible finite input-output segment pairs which are relevant to the level of interest are included. At a given point in time only a subset D of R is known and constitutes the available data about the system at that time.

The base model M is a mathematical structure (e.g., an automaton) and represents the internal structure of the real system as it would appear at the state space level of interest--molecular, in the case of the cell, species in the ecosystem case. The base model is taken to be completely specified and to generate exactly the set R as is its input-output behavior. Since at any point in time only the data D is known, the true base model can be determined only up to a class of models which share the same state space structure and whose presently known input-output behavior is consistent with D.

The *abstract* or *lumped* model M' is also a mathematical structure which represents the present hypotheses concerning the structure of the real system (viz. the base model M). We may identify M' as the abstract system theoretic characterization of the simulation program being executed by the *computer*. This characterization allows us to distinguish the structure of the program relevant to modelling the real system, from the irrelevant computational mechanics employed to carry out its realization.

For any given base model M there are many possible lumped models M' which can correspond to it. The above mentioned simulation models of Weinberg[1] and Patten are examples. In a valid simulation, M' must be connected to M through a preservation relation which ranges from behavioral agreement (validity) at its weakest to some form of internal structural correspondence at its strongest.

The above provides the basis of an approach to the gap between the structure of interest to the modeller (the base model) and what can be simulated (the lumped model). I think that the practical problems associated with this gap can be usefully recast as theoretical questions concerning the relations obtaining among the real system and its base and lumped models, all in the context of the computer. Some of the mathematical tools and concepts for answering these questions are at hand[2] others will have to be developed. The development of these tools and concepts will no doubt be hastened by consciously initiated cooperative efforts on the part of the users (modeller-simulators) and the various system scientists--computer, control, etc. (Each of these has something special to offer which presently may well be inaccessable to others due to the in house dialects which have arisen in these areas.)

In this paper I shall illustrate the use of one of these tools arising from the structure morphism concept developed in (Zeigler, 1972a). The application is to the construction of lumped models from base models in the particular context of brain simulation models. Many other issues arise from the formulation given above, some of which will be touched upon in the conclusion to this paper.

Brain Simulations

The enormous numbers of neurons found in brain tissue makes neuron by neuron simulation of intelligent behavior a practical impossibility. One solution to this problem is

---

[1]For recent results obtained with this model see (Goodman, 1972).

[2]To mention a few which are relevant: realization theory (Kalman, et al., 1969), aggregation theory (Ijiri, 1971; Aoki, 1968), identification theory (Gill, 1965; Balakrishnan, 1969), general systems theory (Klir, 1969, 1972).

to simulate neural nets of small size (about 500 neurons, say). In these computer simulations, the neuron model is supposed to represent salient features of real neurons (Finley, 1967, is a good example; Walker, 1966 and Kauffman, 1969, are related approaches). However one drawback to this approach is that the behavior generated by such small nets may not be very informative about the behavior of nets consisting of 10^{10} neurons (a reasonable estimate of the number of neurons in a human cortex). Another drawback is the difficulty in correlating the behavior of such a neural net with observable behaviors of man or animal as they are expressed in real world actions.

In order to achieve representability and interpretability, many workers have abandoned the neuron as a component in simulation models. Models of cognitive processes (e.g., Feigenbaum, 1963, Newell, et al., 1963) use functionally specified components (e.g., symbol matching subroutines) and make no attempt to correlate these components with a neural substrate.

At a somewhat less removed level one may attempt to identify the components of a model with anatomically significant structures. For example, the optic lobes of an octopus brain may be used as model components (Clymer, 1972, Kilmer, 1969 and Didday, 1970 are best placed in this class).

Finally, one may construct models whose components are intended to represent large groupings of neurons with the difference that both the behavioral and connectivity properties of these components are considered to be directly related to their underlying neural counterparts. This is in contradistinction with the models above in which the interconnection of components may be constrained by known anatomical connectivity but the behavior of a component may not be given a justification in terms of neuron properties. For example, Mortimer's (1970) simulation of the cerebellum employs components that are intended to represent squares of a planar sheet of cortical neurons. The state variables associated with a square are intended to represent the mean firing frequencies of different types of neurons found in the square.

We see then that simulation models of the brain fall in a behavior-structure continuum ranging from the information processing models at the relatively behavioral end to the neuron net models at the relatively structural end.[3] Such a continuum may be characterized formally with the help of function and structure morphisms of various strengths. In this paper, I want to consider the relatively structural end, with the base model being at the neuron net level and the lumped model at the level occupied by Mortimer's simulation. The strong structure morphism concept will be shown to provide a technique for constructing lumped neural nets which are related both behaviorally and structurally (in the sense of the interconnection of components) to the base model.

We shall also look at the following related issue. Often in simulations (e.g., Mortimer, 1970) and in attempts at closed form analysis (e.g., Grossberg, 1971 and MacGregor, 1971) linearly interacting components are assumed. Since it is generally accepted that the neuron is a nonlinear element it is interesting to ask what are conditions under which aggregation of base model neurons can result in linear or bilinear lumped models.

Basic Formalism

The *base model* is a network of Caianiello type (1961) neurons. Each neuron is modelled by a (Moore) automaton:

$$M_\beta = \langle X_\beta, Q_\beta, Y_\beta, \delta_\beta, \lambda_\beta \rangle$$

where $X_\beta = \{0,1\}^n$, $Y_\beta = \{0,1\}$, $Q_\beta = \{0,1\}$ and

$$\delta_\beta(q_\beta, x_1, \ldots, x_n) = 1 \iff \sum_{i=1}^{n} s_i x_i \geq T_\beta$$

$$= 0 \quad \text{otherwise},$$

$$\lambda_\beta(q_\beta) = q_\beta.$$

That is, a neuron β fires at time t+1 ($q_\beta(t+1) = 1$), if and only if the weighted sum of the firing pattern of its neighbors at time t (represented by $(x_1(t), \ldots, x_n(t))$) exceeds threshold T_β.

Let D be the set of all neurons. A network is obtained as a composition of these neurons which is specified by

---

[3]The similarity in this categorization of such models with that of Biondi's and Schmid's paper in this volume is to be noted.

a family of subsets $\{I_\beta | \beta \in D\}$ and a family of maps $\{Z_\beta : \beta \in D\}$ where $Z_\beta : \underset{\alpha \in I_\beta}{X} Y_\alpha \to X_\beta$ determines the input symbol (in X_β) given to neuron β as a function of the symbols appearing at the outputs of its *neighbors* I_β. The simplest assumption for Z_β is an identity map $(Z_\beta(y_1, y_2, \ldots, y_n) = (y_1, y_2, \ldots, y_n))$ so that an output line of a neighbor of β is identified with an input line of β (in a one-one correspondence).

The result of such a composition is a base model $M = \langle Q, \delta \rangle$ where $Q = \underset{\beta \in D}{X} Q_\beta$ and $\delta : Q \to Q$ is specified by a set of *local* transition functions $\{\delta_\beta | \beta \in D\}$ where each δ_β is a slightly disguised form of the neuron transition function above (also called δ_β). This form of representation is called a *structured* automaton, the structure being specified by the sets D, $\{Q_\beta | \beta \in D\}$, $\{I_\beta | \beta \in D\}$, and $\{\delta_\beta | \beta \in D\}$ (in this interpretation D is called a set of coordinates).

The *lumped* model is also a structured automaton $M' = \langle Q', \delta' \rangle$ which is structured by the sets D', $\{Q_{\beta'} | \beta' \in D'\}$, $\{I'_{\beta'} | \beta' \in D'\}$ and $\{\delta'_{\beta'} | \beta' \in D'\}$.

Recall that M' is the object which can actually be simulated while M is the parent system postulated to underly the construction of M' (but which itself can not be simulated because of its extreme complexity).

The basic idea involves considering a pair of onto maps (h,d). $d : D \to D'$ is called the *coordinate mapping* and determines (in our case) the partitioning of neurons into populations or pools, i.e., $(\beta') = d^{-1}(\beta') = \{\beta | d(\beta) = \beta'\}$ is the pool of neurons of the base model which will be grouped together and identified with the coordinate β' in the lumped model. $h : Q \to Q'$ is called an *aggregation* and is structured by a family of onto maps $\{h_{\beta'} | \beta' \in D'\}$ where each $h_{\beta'} : \underset{\beta \in (\beta')}{X} Q_\beta \to Q_{\beta'}$. Note that $\underset{\beta \in (\beta')}{X} Q_\beta$ is the state space of the pool (β') so that each $h_{\beta'}$ represents a *local* aggregation of states and h is the resultant *global* aggregation.

(h,d) is called a *weak structure preserving morphism* if h is a homomorphism of M to M', i.e., for each $q \in Q$,

$h(\delta(q)) = \delta'(h(q))$.

A *strong structure morphism* is a weak structure morphism which also preserves the connection diagram of the base model in the following sense: $\alpha' \varepsilon I'_{\beta'}$ only if there are $\alpha \varepsilon (\alpha')$ and $\beta \varepsilon (\beta')$ and $\alpha \varepsilon I_\beta$ (if α' sends an input to β' in the lumped model then there is at least one neuron α in the (α') pool which sends an input to some neuron β in the (β') pool).

To show how lumped models can be derived from base models using the structure morphism concept, we suppose we have a coordinate mapping $d:D \to D'$ hence a set of pools $\{(\beta') | \beta' \varepsilon D'\}$. Now suppose also that we have selected the input sets $\{I'_{\beta'} | \beta' \varepsilon D'\}$. For a strong structure morphism to hold we require that the pools $\{(\alpha') | \alpha' \varepsilon I'_{\beta'}\}$ send inputs to pool (β') (Fig. 2).

One class of base models is obtained with the following assumptions:

<u>Base Model Class I</u>

1. All neurons $\beta \varepsilon (\beta')$ are identical except for the threshold T_β.
2. Each $\beta \varepsilon (\beta')$ receives $m_{\alpha'}$ inputs from pool (α') for each $\alpha' \varepsilon I'_{\beta'}$ and this accounts for all the inputs to each β (in other words, $m_{\alpha'} = |I_\beta \cap (\alpha')|$ and $I_\beta \subseteq \bigcup_{\alpha' \varepsilon I'_{\beta'}} \{(\alpha')\}$); each of the inputs from pool (α') has the same synaptic weight $s_{\alpha'}$.
3. The $m_{\alpha'}$ input neighbors of $\beta \varepsilon (\beta')$ are distributed in such a way that if $p_{\alpha'}(t)$ is the proportion of neurons in (α') firing at time t, then $m_{\alpha'} p_{\alpha'}(t)$ is the number of pulses (1 s) appearing on the input lines at time t. Thus β will fire at time t+1 if and only if

$$\sum_{\alpha' \varepsilon I'_{\beta'}} m_{\alpha'} s_{\alpha'} p_{\alpha'}(t) \geq T_\beta .$$

4. T_β has a distribution Φ in (β'), i.e., $\Phi(x)$ = (the proportion of β in (β') for which $T_\beta \leq x$)

$$= \frac{|\{\beta | \beta \varepsilon (\beta') \text{ and } T_\beta \leq x\}|}{|(\beta')|} .$$

From 3. and 4. the proportion of β in (β') that fire at time t+1,

$$p_{\beta'}(t+1) = \Phi(\sum_{\alpha' \varepsilon I'_{\beta'}} m_{\alpha'} s_{\alpha'} p_{\alpha'}(t)).$$

Formalizing this result gives us the structure of the lumped model $M' = \langle Q', \delta' \rangle$. We choose $Q'_{\beta'} = [0,1]$ for each $\beta' \varepsilon D'$ and having already chosen the family $\{I'_{\beta'} | \beta' \varepsilon D'\}$ we structure δ' by the family $\{\delta'_{\beta'} | \beta' \varepsilon D'\}$ where $\delta'_{\beta'}: \mathop{X}_{\alpha' \varepsilon I'_{\beta'}} Q'_{\alpha'} \to Q'_{\beta'}$ is given by

$$\delta'_{\beta'}(p_{\alpha'_1}, p_{\alpha'_2}, \ldots, p_{\alpha'_{|I'_{\beta'}|}}) = \Phi(\sum_{\alpha' \varepsilon I'_{\beta'}} m_{\alpha'} s_{\alpha'} p_{\alpha'})$$

for all $(p_{\alpha'_1}, p_{\alpha'_2}, \ldots, p_{\alpha'_{|I'_{\beta'}|}}) \; \varepsilon \mathop{X}_{\alpha' \varepsilon I'_{\beta'}} Q'_{\alpha'}$.

To show that M' is a strong structure morphic image of M we must come up with an aggregation $h: Q \to Q'$ and show this to be a homomorphism. Since the coordinates of the lumped model have an interpretation as firing proportions of pools it is natural that our choice of $\{h_{\beta'} | \beta' \varepsilon D'\}$ reflects this. Thus let $h_{\beta'}: \mathop{X}_{\beta \varepsilon (\beta')} Q_\beta \to Q'_{\beta'}$ be given by

$$h_{\beta'}(q_{\beta_1}, q_{\beta_2}, \ldots, q_{\beta_{|(\beta')|}}) = \frac{\sum_{\beta \varepsilon (\beta')} \lambda_\beta(q_\beta)}{|(\beta')|}$$

$$= \frac{\sum_{\beta \varepsilon (\beta')} q_\beta}{|(\beta')|}$$

for all $(q_{\beta_1}, q_{\beta_2}, \ldots, q_{\beta_{|(\beta')|}}) \; \varepsilon \mathop{X}_{\beta \varepsilon (\beta')} Q_\beta$.

Interpretatively, $h_{\beta'}$ gives the proportion of neurons in (β') which are firing, given the states of all the neurons in pool (β').

It can now be formally verified that h constructed from $\{h_{\beta'}\}$ is a homomorphism. We do this by verifying for each $\beta' \varepsilon D'$ that $(h_{I'_{\beta'}}, h_{\beta'})$ is a function morphism from $\delta_{(\beta')} \circ p_{I_{(\beta')}}$ to $\delta'_{\beta'} \circ p_{I'_{\beta'}}$ (see Proposition 5.1 of Zeigler (1972)).

With Φ taken to a Gaussian with mean μ and standard deviation σ we can investigate the local lumped coordinate transition function:

$$\delta'_{\beta'}(p_1,\ldots,p_n) = \Phi(\sum_{i=1}^{n} m_i s_i p_i).$$

Note that $\sum_{i=1}^{n} m_i s_i^- \leq \sum_{i=1}^{n} m_i s_i p_i \leq \sum_{i=1}^{n} m_i s_i^+$ where

$s_i^- = \min[s_i,0]$ and $s_i^+ = \max[s_i,0]$. So with $x = \sum_{i=1}^{n} m_i s_i p_i$ we take x to lie in the interval $[-1,1]$.

With $\phi(x) = \frac{1}{\sqrt{2\pi}} \int_{\infty}^{x} e^{-x^2} dx$, we have $\Phi(x) = \phi(\frac{x-\mu}{\sigma})$.

The shape of $\Phi(x)$ in the region $x \in [-1,1]$ will depend on μ and σ (Fig. 3). With σ small we obtain a sharp S-type curve. In this condition all the neurons tend to fire or not in unison so that the aggregate behavior approximates the behavior of a representative neuron. Thus we can provide conditions under which this commonly employed assumption can hold.

At the other extreme, when σ is large, $\Phi(x)$ is approximately linear in $[-1,1]$. In this condition the neurons are well spread out in their resistance to firing (here determined by the threshold) so that as the input x is increased the fraction firing increases proportionally . With $\Phi(x) \hat{=} ax+b$ where $a \geq 0$ and $b \geq 0$ we have

$$\delta'_{\beta'}(p_1,\ldots,p_n) = \sum_{i=1}^{n} am_i s_i p_i + b$$

so that with $b = 0$ the local transition function is linear as often assumed. But note that for $b > 0$ the system is actually nonlinear. Moreover it appears that we cannot choose (σ,μ) such that b is small without at the same time drastically reducing the slope a. However, the system is truly linear when the state is measured relative to an equilibrium level q^*, (i.e., if $\delta'(q) = Aq+B$ is the global transition function, an equilibrium level q^* is one for which $\delta(q^*) = q^*$). The interesting case is where $q^* \in [0,1]^n$, i.e., represents a possible state for the aggregative network which we can identify as the background firing pattern.

Base Model Class II

The following assumptions yield quite different results. Instead of assumption 1. We use

1'. All neurons $\beta \varepsilon (\beta')$ are identical except for a synaptic gain g_β such that

$$\delta(q,x_1,\ldots,x_n) = 1 \text{ if } g_\beta \Sigma s_i x_i \geq T.$$
$$= 0 \text{ otherwise.}$$

Instead of 4. assume

4'. $\Phi(x)$ = the proportion of $\beta \varepsilon (\beta')$ such that $g_\beta \leqslant x$.

i.e.,
$$\Phi(x) = \frac{|\{\beta | x \geqslant g_\beta\}|}{|(\beta')|} .$$

Then the proportion of $\beta \varepsilon (\beta')$ which will fire at time t+1 is given by

$$\delta'_{\beta'}(p_1,\ldots,p_n) = 1 - \Phi(T/\Sigma m_i s_i p_i) \text{ for } \Sigma m_i s_i p_i \geqslant 0$$
$$= \Phi(T/\Sigma m_i s_i p_i) \text{ for } \Sigma m_i s_i p_i < 0$$

Fig. 4 demonstrates that for the same values of σ and μ we obtain much more nonlinearity than previously. Thus to account for possible linear aggregative structure we would have to reject the second set of assumptions, i.e., it is more likely that thresholds rather than synaptic gains are the variable elements of neurons in pools.

Refinements of the Base Model

The foregoing has illustrated the use of structure morphism to derive lumped models from base models. The base model neurons were rather simple however and the reader may ask whether this process can be carried out when more realistic base models are considered. In answer to this question, I have applied the same technique to base models of noisy neurons with refractory behavior. The basic idea will be quickly sketched and more details can be found in the Appendix.

The base model neurons now take the form:

$$M_\beta = \langle X_\beta, Q_\beta, Y_\beta, \delta_\beta, \lambda_\beta \rangle$$

where $X_\beta = \{0,1\}^n$, $Y_\beta = \{0,1\}$, $Q_\beta = N = \{0,1,2,\ldots\}$

and $\delta_\beta(r,(x_1,x_2,\ldots,x_n)) = 0 \iff \sum_{i=1}^{n} s_i x_i \geq T_\beta(r)$

$= r+1$ otherwise;

and $\lambda_\beta(r) = 1 \iff r = 0$

$= 0$ otherwise.

Thus the state r records the elapsed time since the neuron last fired. Whether a neuron fires ($r = 0$) or not depends on its total excitation ($\sum_{i=1}^{n} s_i x_i$) and its recovery stage given by the recovery characteristic $T_\beta : N \to R$ (the real numbers) which is typically a decreasing function.

We make the same assumptions 1.,2.,3. as before. Instead of 4. we assume

4'. With each state r we can associate a distribution Φ_r where, of those neurons $\beta \in (\beta')$ which are in state r, $\Phi_r(x)$ is the proportion for which $x \geq T_\beta(r)$. More formally if

$(\beta')^q_r = \{\beta \mid \beta \in (\beta'),\ proj_\beta(q) = r\}$ then

$$\Phi_r(x) = \frac{|\{\beta \mid \beta \in (\beta')^q_r,\ x \geq T_\beta(r)\}|}{|(\beta')^q_r|} .$$

Note that $(\beta')^q_r$ (= the number of neurons in pool (β') which are in state r when the global state is q) is dependent on q and our assumption 4'. is essentially that this dependence is not transmitted to Φ_r, i.e., $\Phi_r(x)$ is the same no matter at which global state q or time t it is called upon.

4'. is evidently a much stronger assumption than is 4. To provide a possible realization of 4'. we suppose that Φ_r is a normal distribution with mean T(r) and standard deviation σ, i.e.,

$$\Phi_r(x) = \phi\left(\frac{x - T(r)}{\sigma}\right).$$

This can be derived from the assumption that for each $\beta \in (\beta')$, $T_\beta(r) = T(r) + t_\beta$, where t_β is normally distributed in (β') with mean 0 and standard deviation σ. This distribution can be maintained by adding a noise source to the base model neuron (corresponding to chemical influences not accounted for in the deterministic model).

The lumped model M' obtained by an appropriate strucsure morphism may be analyzed using Markov chain concepts. Under certain conditions, the local transition functions take the form of first order systems with time constant dependence on input. With $p_{\beta'}$ representing proportion of neurons in pool (β') which fire, and $x = \sum'_{\alpha' \varepsilon I'_{\beta'}} m_{\alpha'} s_{\alpha'} p_{\alpha'}$ we obtain

$$\delta'_{\beta'}(p_{\beta'}, x) = \rho_x(p_{\beta'} - \pi_x) + \pi_x$$

where under certain conditions the equilibrium firing rate π_x is a linear function of x and the time constant ρ_x has the dependence on x shown in Fig. 5. As suggested by R. Mohler, this can be viewed as a bilinear system with bang-bang control. To my knowledge, neuron net models with these kinds of components have not been heretofore considered by brain modellers.

Conclusions

Different base model assumptions may yield lumped models with distinctive characteristics, (cf., base model classes I, II and the refractory extension of class I).

It is clear from the assumptions 1 through 4 that the partition of neurons into pools is not arbitrary. In fact it is unlikely that very many partitions will meet the special requirements of 2 and 3. Thus while in the abstract, the requirement of strong structure morphism may provide no practical guide for selecting coordinate partitions, in particular applications, guided by intuition it may isolate a reasonably small class of possibilities.

One may not be able to test the base assumptions directly. Thus while assumption 2 may be confirmed or disconfirmed by anatomical connectivity evidence, this is generally not true for assumption 3. The latter requires that each neuron in (β') obtain a precise reading of the firing frequencies of each input pool (α'). Anatomical evidence will suffice if it reveals that each neuron in (β') receives one input from each of the neurons in an input pool. Thus every neuron reads exactly the proportion of firing neurons in that pool. However if the inputs from a pool are relatively few in number it may or may not be that each terminal neuron [in (β')] samples a representative set of input pool neurons. The only way to verify this is to obtain the spacio-temporal firing pattern of a pool. Because of the large numbers involved this may be impossible without question begging assumptions. (This is the general problem of state observability.)

Base model assumptions are subject to indirect confirmation. The chain of induction runs as follows:

The internal structure of the lumped model while not directly testable is confirmed whenever the model accounts for the observable system behavior. Zeigler (1972a) gives conditions under which the internal structure of such a validated model may be reliably inferred to reflect that of the base model. When such a structure morphism is confirmed the local conditions assumed in deriving the lumped model structure from the base model are met. Thus a set of base assumptions which yields a lumped model which is confirmed is itself confirmed. A set of assumptions which results in an incompatible lumped model must be rejected.

For example, as already indicated should a linear lumped model (linearity being a nondirectly observable characteristic) be validated over a large number of behavioral experiments we would be in a position to distinguish between the assumptions distinguishing classes I and II.

Thus identification of the aggregated model permits identification of the base model (though of course the former is able to proceed much more slowly than can the latter).

We can see explicitly how many distinct base models may be mapped into one lumped model. For example, while $m_{\alpha'}$ and $s_{\alpha'}$ are distinct parameters for the base model only their product $m_{\alpha'}s_{\alpha'}$ appears as a parameter in the lumped model. Again there may be many possible distinct base models which satisfy assumptions 2., 3., and 4. Thus the efficacy of aggregation in reducing the number of to-be-specified parameters is apparent. This of course means that many fewer models need be tried out in order to validate or invalidate the lumped model class.

The reduction in state space complexity afforded by aggregation is evident, this being 2^{nm} versus $(n+1)^m$ where n is the number of neurons in a pool and m is the number of pools.

Acknowledgements

The approach of Stephen Kaplan to scientific modelling in general, and brain modelling in particular, had much to do with suggesting the formal development taken here.

Appendix

I shall outline the analysis of the lumped model in the case that the base model consists of noisy refractory neurons.

The first thing to be done is to assign an appropriate state space to each pool of neurons. It is natural to take a state $p_{\beta'}$ of pool (β') to be the distribution of its neurons in the various recovery states. Thus let $p_{\beta'}(t) = (p^0_{\beta'}(t), p^1_{\beta'}(t), \ldots, p^r_{\beta'}(t), \ldots)$ where $p^r_{\beta'}(t)$ is the proportion of neurons β in (β') which are in recovery state r at time t. Thus, $p^0_{\beta'}(t)$ is the proportion which are firing at time t and $\sum_{r=0}^{\infty} p^r_{\beta'}(t) = 1$. From the assumed base model component we see that the neurons in state r at time t can go only two routes - one is to go to state r+1 at time t+1, the other is to go to state 0 at time t+1. Thus we have

$$p^{r+1}_{\beta'}(t+1) = [1 - \Phi_r(\sum_{i=1}^{|I'_{\beta'}|} m_i s_i p^0_{\alpha'_i}(t))]p^r_{\beta'}(t)$$

for $r \varepsilon N$ and

$$p^0_{\beta'}(t+1) = \sum_{r=0}^{\infty} \Phi_r(\sum_{i=1}^{|I'_{\beta'}|} m_i s_i p^0_{\alpha'_i}(t))p^r_{\beta'}(t)$$

This suggests we consider the lumped model $M' = <Q', \delta'>$ where $Q'_{\beta'} = \{p | p = (p^0, p^1, \ldots, p^j, \ldots), p_j \varepsilon [0,1]$ for each $j \varepsilon N$ and $\sum_{j \varepsilon N} p_j = 1\}$,

$Q' = \underset{\beta' \varepsilon D'}{X} Q'_{\beta'}$ and $\delta'_{\beta'} : \underset{\alpha' \varepsilon I'_{\beta'}}{X} Q'_{\alpha'} \rightarrow Q_{\beta'}$ is given by

$$\delta'_{\beta'}(p_1, p_2, \ldots, p_{|I'_{\beta'}|}) = (\sum_{r=0}^{\infty} \Phi_r(\sum_{i=1}^{|I'_{\beta'}|} m_i s_i p^0_i)\ p^r_1,$$

$$1 - \Phi_0(\sum_{i=1}^{|I'_{\beta'}|} m_i s_i p^0_i) p^0_1, \ldots, 1 - \Phi_j(\sum_{i=1}^{|I'_{\beta'}|} m_i s_i p^0_i) p^j_1, \ldots)$$

where $p_i = (p_i^0, p_i^1, p_i^2, \ldots, p_i^j, \ldots)$ for $i=1,2,\ldots,|I'_\beta|$.

It is more revealing to consider M' as a composition of machines $\{M'_{\beta'} | \beta' \in D'\}$ where each $M'_{\beta'} = \langle X'_{\beta'}, Q'_{\beta'}, Y'_{\beta'}, \delta'_{\beta'}, \lambda'_{\beta'} \rangle$ is given by $X'_{\beta'} = R$ and $Y'_{\beta'} = [0,1]$; $\delta'_{\beta'} : Q'_{\beta'} \times X'_{\beta'} \to Q'_{\beta'}$ is determined for each $x \in X'_{\beta'}$ by the matrix

$$P(x) = \begin{bmatrix} \Phi_0(x) & 1-\Phi_0(x) & 0 & 0 & \cdot & \cdot & \cdot \\ \Phi_1(x) & 0 & 1-\Phi_1(x) & 0 & \cdot & \cdot & \cdot \\ \Phi_2(x) & 0 & 0 & 1-\Phi_2(x) & \cdot & \cdot & \cdot \\ \vdots & \vdots & \vdots & \vdots & \vdots & \vdots & \vdots \end{bmatrix}$$

such that $\delta'_{\beta'}(p,x) = pP(x)$; $\lambda'_{\beta'} : Q'_{\beta'} \to Y'_{\beta'}$ is the projection of p on the 0 coordinate hence may be determined by the vector

$$\Lambda = \begin{bmatrix} 1 \\ 0 \\ 0 \\ \vdots \end{bmatrix}$$

so that $\lambda'_{\beta'}(p) = p\Lambda$.

M' as a composition of the $\{M'_{\beta'} | \beta' \in D'\}$ is then obtained with the connecting maps $\{Z'_{\beta'} | \beta' \in D'\}$ where $Z'_{\beta'} : \underset{\alpha' \in I'_{\beta'}}{X} Y'_{\alpha'} \to X'_{\beta'}$ is given by

$$Z'_{\beta'}(y_1, y_2, \ldots, y_{|I'_{\beta'}|}) = \sum_{i=1}^{|I'_{\beta'}|} m_i s_i y_i .$$

Note that with x fixed, P(x) formally represents a Markov chain. However our *interpretation* of the element p is that of a *state* vector (giving the proportions $\{p_i\}_{i=0}^{\infty}$ of neurons which are in the recovery states $i \in N$) *not* a probability distribution. For the corresponding

Markov chain, N is the state set not a subset of $[0,1]^N$. With x allowed to vary the formalism is that of (Moore) stochastic automata (Paz, 1970) again with the differing interpretation concerning what is the state of the system and consequently what is the domain of the output map ($\lambda'_{\beta'}$ maps a subset of $[0,1]^N$ to $[0,1]$; we could not obtain the desired effect (of outputting the firing proportion p^0) were we to use the stochastic automaton formalism $\lambda'_{\beta'}$ mapping $N \to [0,1]$.)

Equilibrium firing distributions

The formal similarity however enables us to apply those theorems of Markov chains which are consistent with our interpretation.

Thus from Parzen (1962) we know that under certain conditions there exists an equilibrium state Π^x (called a stationary distribution) such that $\Pi^x = \Pi^x P(x)$. Moreover under more stringent conditions this state is globally stable (hence called a long run distribution) so that for all $p \in Q'_{\beta'}$, $\lim_{t\to\infty} pP^t(x) = \Pi^x$. We interpret this to mean that if the input to a pool of neurons is held constant, an equilibrium distribution of the proportions of neurons in the various states will be always reached, and which, in particular the proportion that fire at every time is constant $p_0(t) = \Pi_0^x$, (here $p(t) = pP^t(x)$).

Fig. 6 plots Π_0^x against x in the region $0 \le x \le 1$ where

$$\Phi_r(x) = \phi\left(\frac{x-T_\mu(r)}{\sigma}\right) \text{ and } T_\mu(r) = u\left(1-\frac{r}{20}\right) \text{ for } 0 \le r \le 20$$
$$= 0 \text{ for } r > 20.$$

The case $\mu = 0$ reduces to that of the nonrefractory neuron. As μ is increased the curve tends to become more linear until finally at $u = 2$ a rather linear response is observed though at a relatively low output level. Note that a comparatively small spread in threshold is assumed ($\sigma = \frac{1}{5}$) so that the effects observed are not solely that of the threshold distribution. The comparable curves of Fig. 2 superimposed in Fig. 6 show this clearly. The effect of the refractory behavior modelled by T(r) is to spread out the firing of the neurons in a pool in time thus augmenting the spread of firing in "space" modelled by the distribution of the fixed component t_β.

Of course, this can only be achieved at a cost in latency of response and it is to this that we now turn.

Time Required to Achieve Equilibrium

Supposing that a globally stable state Π^x exists there are conditions (Parzen, 1962) under which this state is always approached exponentially, i.e., there are numbers ρ_x and M_x such that $0 \leq \rho_x < 1$ and $0 \leq M_x < \infty$ and for all $t = 0,1,2,\ldots$ $|p_{j,0}(t) - \Pi_x^0| \leq M_x\rho_x^t$ where $p_{j,0}(t)$ is the $(j,0)$th entry in $P^t(x)$. From this and assuming that the neuron state set is finite, i.e., $Q_\beta = \{0,1,2,\ldots,n-1\}$ we obtain

$$|p^0(t) - \Pi_x^0| \leq nM_x\rho_x^t$$

for all $p \in [0,1]^n$.

Since ρ_x is explicitly calculable from $P(x)$ this suggests that we might try to reduce the lumped model components to first order systems which would realize the exponential convergence but disregard possible oscillation, viz.

$$p^0(t+1) = \rho_x(p^0(t) - \Pi_x^0) + \Pi_x^0.$$

Explicitly, we would have $M'_{\beta'} = \langle X'_\beta, Q'_\beta, Y'_\beta, \delta'_{\beta'}, \lambda'_{\beta'}\rangle$ where $X'_{\beta'} = R$, $Q'_{\beta'} = [0,1]$, $Y'_{\beta'} = [0,1]$ and for each $x \in X'_{\beta'}$ for which a Π_x^0 exists we have

$$\delta'_{\beta'}(p^0,x) = \rho_x(p^0 - \Pi_x^0) + \Pi_x^0$$

and

$$\lambda(p^0) = p^0.$$

Let us suppose that Π_x^0 exists for every $x \in X'_{\beta'}$ so that $\delta'_{\beta'}$ is totally defined. The lumped moded $M' = \langle Q', \delta'\rangle$ then is a composition of the first order system $\{M'_{\beta'} | \beta' \in D'\}$.

Formal Noise Model

To find a plausible basis for the assumption that Φ_r is independent of state q we shall explicitly account for the noise source. To do this we expand the state of a base model neuron to include a pseudo-random number generator. Let $M_\beta = \langle Q_\beta, X_\beta, Y_\beta, \delta_\beta, \lambda_\beta \rangle$ where

$$Q_\beta = N \times R,\ X_\beta = \{0,1\}^n,\ Y_\beta = \{0,1\}$$

and

$$\begin{aligned}\delta_\beta(j,t,(x_1 x_2,\ldots,x_n)) &= (0,G(t)) \iff \sum_{i=1}^{n} s_i x_i \geq T(j)+t \\ &= (j+1,G(t)) \quad \text{otherwise.}\end{aligned}$$

Here $G{:}R \to R$ generates the random sequence $t = G^0(t), G^1(t), G^2(t), G^3(t), \ldots, G^n(t), \ldots$ which is a realization of a stochastic process for which $\{G^n | n = 0,1,2,\ldots\}$ are independent identically distributed normal random variables with mean 0 and standard deviation σ ($G^1 = G$ and $G^{n+1} = G \circ G^n$). As before

$$\begin{aligned}\lambda_\beta(j,t) &= 1 \iff j=0 \\ &= 0 \quad \text{otherwise.}\end{aligned}$$

For this model, the probability that a neuron in state j has a threshold not greater than x, is given by

$$\begin{aligned}P_r\{T_\beta \leq x\} &= P_r\{T(j)+t_\beta \leq x\} \\ &= P_r\{T_\beta \leq x-T(j)\} \\ &= \phi\left(\frac{x-T(j)}{\sigma}\right).\end{aligned}$$

By the strong law of large numbers, the proportion of neurons in state j having $t_\beta \leq x$ closely approximates $P_r\{T_\beta \leq x\}$ provided the number of such neurons is sufficiently high. Thus for $q \in \underset{\beta\in D}{X} Q_\beta = (N \times R)^{|D|}$ and $|(\beta')^q_j|$ sufficiently large,

$$\Phi_j(x) = \frac{|\{\beta\varepsilon(\beta')^q_j | proj_2(proj_\beta(q)) \leq x-T(j)\}|}{|(\beta')^q_j|}$$

$$= \phi\left(\frac{x-T(j)}{\sigma}\right)$$

This is reminiscent of the deterministic model approximations to stochastic population genetic models. From the approach here it is possible to point to the conditions justifying the jump from stochastic to deterministic models.

References

1. Aoki, M. (1969) "Control of Large-Scale Dynamic Systems by Aggregation" IEEE Trans. Aut. Cont., AC-13, 3.

2. Balakrishnan, A.V. (1969) "Identification in Automatic Control Systems" Automatica, 5.

3. Caianiello, E.R. (1961) J. of Theor. Biol., 2, 204.

4. Clymer, J.C. (1972) Doctoral Thesis Prospectus, Department of Computer and Communication Sciences, The University of Michigan.

5. Didday, R.L. (1970) "The Simulation and Modeling of Distributed Information Processing in the Frog Visual System" Technical Report 6112-1, Information Systems Lab., Stanford University.

6. Feigenbaum, E.A. (1963) "The Simulation of Verbal Learning Behavior" In Computers and Thought, (E.A. Feigenbaum and J. Feldman, eds.).

7. Finley, M. (1967) "An Experimental Study of the Formation and Development of Hebbian Cell-Assemblies by Means of a Neural Network Simulation" Department of Computer and Communication Sciences, The University of Michigan, Doctoral Thesis.

8. Gill, A. (1965) Introduction to the Theory of Finite State Machines, McGraw-Hill, New York.

9. Goodman, E.D. (1972) "Adaptive Behavior of Simulated Bacterial Cells Subjected to Nutritional Shifts" Department of Computer and Communication Sciences, The

University of Michigan, Doctoral Thesis.

10. Grossberg, S. (1971) "On the Dynamics of Operant Conditioning" J. of Theor. Biol., 33, 2.

11. Ijiri, Y. (1971) "Fundamental Queries in Aggregation Theory" J. Amer. Stat. Assoc., 66, 336.

12. Kalman, R.; M. Arbib, and P. Falb (1969) Topics in Mathematical Systems Theory, Van Nostrand, New York.

13. Kauffman, S.A. (1969) "Metabolic Stability and Epigenesis in Randomly Constructed Genetic Nets" J. of Theor. Biol., 22, 437-467.

14. Kilmer, W.L.; W.S. McCulloch, and J. Blum (1969) "A Model of the Vertebrate Central Command System" Inter. J. Man-Machine Studies, 1, 279-309.

15. Klir, G. (1969) An Approach to General Systems Theory, Van Nostrand, New York.

16. ___, (1972) Recent Trends in General Systems Theory, Wiley, New York.

17. MacGregor, R.J. (1971) "Intrinsic Oscillations in Neural Networks: A Linear Model for the nth Order Loop" Math. BioSci., 11, 34.

18. Mortimer, J.A. (1970) "A Cellular Model for Mammalian Cerebellar Cortex" Department of Computer and Communication Sciences, The University of Michigan Technical Report No. 03296-7-T.

19. Newell, A. and H.A. Simon (1963) "GPS, A Program that Simulates Human Thought" In Computers and Thought, (E.A. Feigenba,m and J. Feldman, eds.)

20. Parzen, E. (1962) Stochastic Process, Holden Day.

21. Patten, B.C. (1972) A State Space Model for Grassland, (To appear).

22. Paz, A. (1971) Introduction to Probabilistic Automata, Academic Press, New York.

23. Walker, C.C. and W.R. Ashby (1966) "On Temporal Characteristics of Behavior in Certain Complex Systems" Cybernetics, 3, 2, 11-108.

24. Zeigler, B.P. (1971a) "On the Formulation of Problems in Simulation and Modeling in the Framework of Mathematical System Theory" Proceedings of the Sixth International Congress on Cybernetics, Namur, Belgium.

25. ___, (1971b) "Automaton Structure Preserving Morphisms with Applications to Decomposition and Simulation" in Theory of Machines and Computations, 295-309, Academic Press, New York.

26. ___, (1972a) "Towards a Formal Theory of Modeling and Simulation: Structure Preserving Morphisms" (to appear) J. of the Assoc. of Comp. Mach.

27. ___, (1972b) "Modelling and Simulation: Structure Preserving Relations for Continuous and Discrete Time Systems" Computers and Automata, Brooklyn Polytechnic Press.

28. Zeigler, B.P. and R. Weinberg (1970) "System Theoretic Analysis of Models: Computer Simulation of a Living Cell" J. Theor. Biol., 29, 35-56.

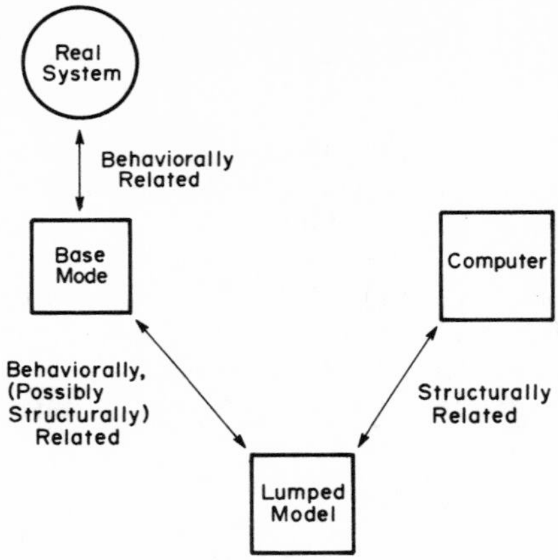

Figure 1: The Basic Elements of a Simulation.

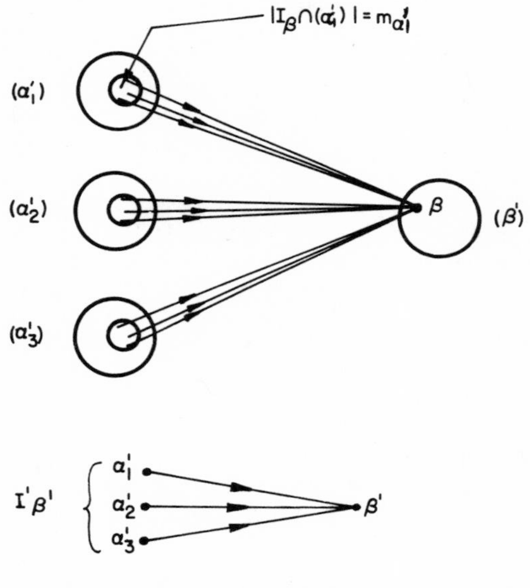

Figure 2: Component β' in the lumped model receives inputs from components α_1', α_2' and α_3'. This reflects the situation in the base model where each neuron β in pool (β') receives $m_{\alpha_i'}$ inputs from pool (α_i') for i = 1,2,3.

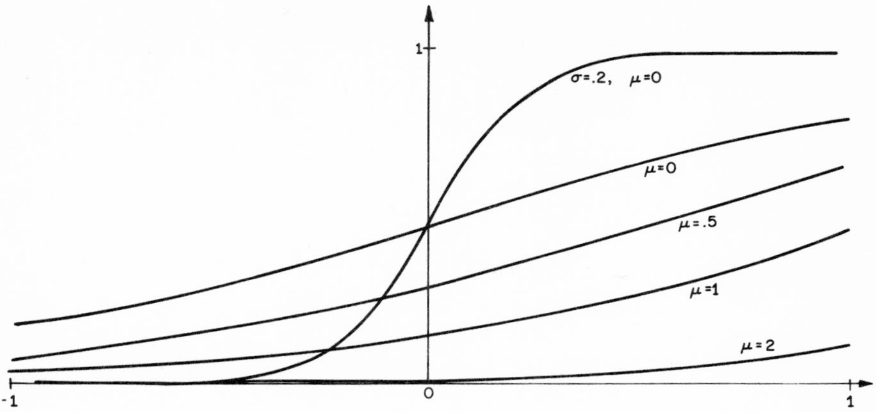

Figure 3: $\phi(\frac{x-\mu}{\sigma})$ versus x for σ = .2 (sharp curve) and σ = 1 with μ = 0,.5,1,2.

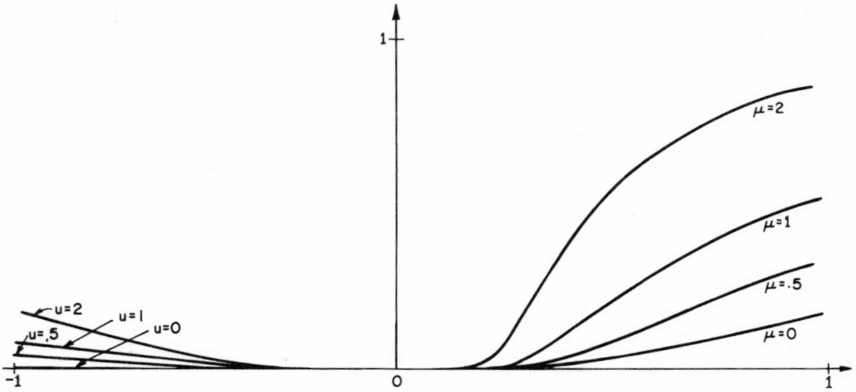

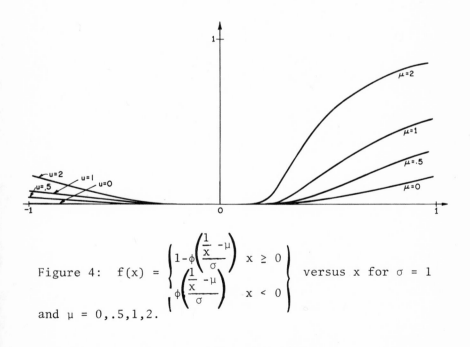

Figure 4: $f(x) = \begin{cases} 1-\phi\left(\dfrac{\frac{1}{x}-\mu}{\sigma}\right) & x \geq 0 \\ \phi\left(\dfrac{\frac{1}{x}-\mu}{\sigma}\right) & x < 0 \end{cases}$ versus x for σ = 1 and μ = 0,.5,1,2.

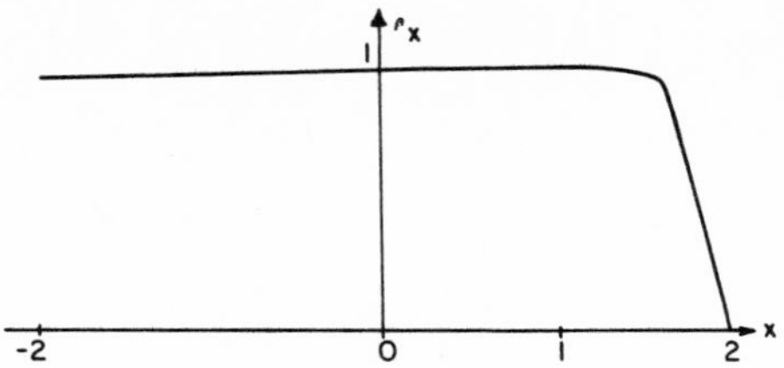

Figure 5: Time constant ρ_x as a function of x.

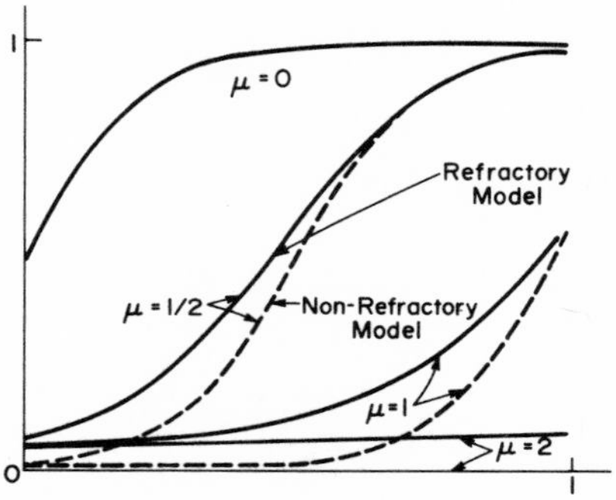

Figure 6a:
The equilibrium distribution Π_0^x plotted against x for the refractory (solid line) and nonrefractory (dotted line) cases.

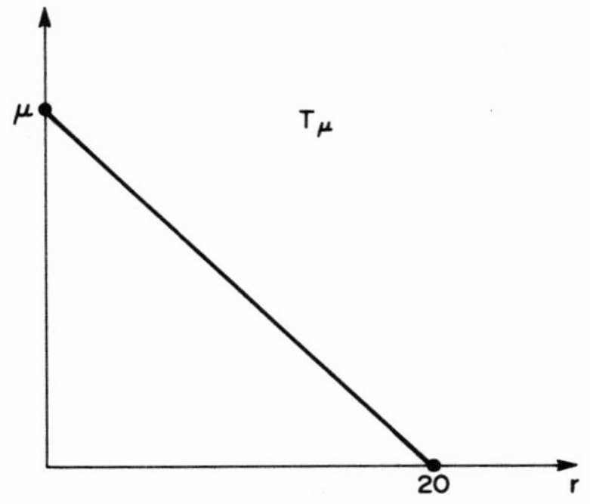

Figure 6b: The threshold curve T_μ with parameter μ used in the computations of Fig. 5 and Fig. 6a.

A DYNAMIC UPPER BOUND OF THE UNCERTAINTY IN LINEAR SYSTEM APPROXIMATION

Francesco Donati

Gruppo di Informatica e Automatica
I. E. N. G. F. - Politecnico di Torino (Italy)

Abstract

The paper is concerned with the evaluation of the uncertainty in linear system approximation. A linear system is given with an approximant model: a method is presented to obtain a simple non linear dynamic system whose output represent an upper bound of the instantaneous error for any admissible input.

Introduction

The control system analyst often finds himself dealing with the problem of designing controls of systems, of which it is not possible to get exact mathematical models. Then, designing the control it is necessary to take into account the uncertainty, which derives from modeling.

The uncertainty can be expressed in many different ways. Three main types of description can be pointed out,

a. uncertainty given by disturbances added to the inputs and to the outputs
b. uncertainty on the model parameter values
c. uncertainty "in norm".

Type a. can be considered as an additive type of

uncertainty, while b. and c. are of parametric type. In b. the model structure (or in other words the model order) is supposed exactly the same as that of the actual system. The error is attributed to an uncertainty in the parameter values which are assumed to range in a given set. In c. the model order is assumed to be lower than that of the actual system. Then, an error is present due to the reduction of the order and the uncertainty cannot be expressed by ranging some model parameters in a set, but it can be done giving a norm of the error system.

The error system E is the one that under the same input u of the actual system S and the model S' gives in output the difference of outputs of S and S'; e = y - y' (fig. 1).

Then, given a set U of admissible inputs and assumed that input and outputs belong to given normed spaces and that E has zero initial conditions, the error system norm is defined (when it exists) by:

$$\|E\| = \sup \frac{\|e\|}{\|u\|}, \quad u \in U, \quad \|u\| > 0$$

Obviously, the knowledge of $\|E\|$ gives a constraint on the uncertainty of the model output, since:

$$\|e\| = \|y - y'\| \leq \|E\| \; \|u\|, \quad u \in U$$

In this note the line of the uncertainty definition by a norm of the error system will be followed.

Review of prior work

The problem of the approximation in norm (or uniform approximation) was particulary developed by the research group of Torino (Italy) as results in references.

In [1] the approximation in norm is considered for linear system when the input and output functions are assumed to belong to subspaces of the Hilbert spaces L^2.

The existence of an optimal model is proved and computer oriented algorithms are given for the evaluation of the error system norm and the optimal model, when the actual system is described by a set of input-output records.

In [2] the general problem of the approximation in norm is considered for linear invariant or time-varying systems modeled by the impulsive response.

Expressions of system norms are given in several cases when different kinds of norm (C, L^1, L^2) are assumed for the input and output functions.

The uniform approximation criterion is applied in [3] in optimal modeling time-variant linear system by finite rank operators. The paper [4] is concerned with the evaluation, from input-output data, of the norm of linear operators acting between Banach spaces. Computer oriented algorithm are developed.

Linear, time-varying, distributed parameters dynamic system are considered in [5]. The problem of uniform approximation is developed when inputs, initial conditions and outputs are assumed to belong to some normed linear spaces.

The problem of the uniform approximation of system when the inputs are sample functions of a stochastic process is considered in [6], [7], several criteria are exposed and the related properties are derived.

In recent years also the control problem of system in presence of uncertainty in norm has been considered and some interesting solution are proposed [8, 9, 10], following a min-max strategy. These papers show as the definition of the uncertainty by the error system norm can be an useful tool in control design.

Expressing the uncertainty between model and system by an error system norm, the main limit of the prior work, may be, consists in the complete loss of the dynamics of the error system. There follows that, sometimes, it is difficult to give a solution to certain control problems, which is not too pessimistic.

In this note the problem of evaluating the uncertainty of a model by a norm of the error system is approached in a different way from the previous works, in order to preserve some aspects of the dynamics of the error system. A non linear dynamic system is derived, whose output gives an instantaneous upper bound of the error between the system and model outputs.

Dynamic upper bound of error system

Let S be a linear dynamic, n-order, stable system and let S' be an approximant, m-order, linear model of S ($m<n$). Let $u(t) \in R_i$ be the i-dimension input vector, the same both for S and for S'. Let $y(t)$, $y'(t) \in R_j$ be the j-dimension output vectors, respectively of S and S'.

The error system, $E = S - S'$, is defined as the linear $(n + m)$ - order system, with input $u(t)$ and output $e(t) = y(t) - y'(t)$.

Let $\| \ \|_u$ be a norm defined on the input space R_i and $\| \ \|_y$ a norm defined on the output space R_j.

We assume, moreover, that the set of acceptable input functions $U(-\infty, +\infty)$ is defined by the property:

$$\int_{-\infty}^{+\infty} (\|u(t)\|_u)^p \, dt < \infty \tag{1}$$

where $p \geq 1$.

The aim of this section is the research of a dynamic system E_M which is an upper bound of the error system E according to the following definition.

Definition

Let E_M be a stable dynamic system, with the same input $u \in U(-\infty, \infty)$ applied to E. Let E_M have one real output $z(t)$. We say that E_M is a dynamic upper bound of E, if it results

$$z(t) \geq \|e(t)\|_y \tag{2}$$

for any time value, $-\infty < t < \infty$ and input function $u \in U(-\infty, +\infty)$.

There holds the following

Theorem 1

Given a linear stable system, with one input $(\|u(t)\|_u)^p$ and one output $(z'(t))^p$:

$$\begin{aligned} x &= Ax - B(\|u(t)\|_u)^p \\ (z'(t))^p &= Cx \end{aligned} \tag{3}$$

there exists a finite value M(t), such that results:

$$z(t) = M(t)\, z'(t) \geq \|e(t)\|_y \quad \text{for any } u \in U(-\infty, +\infty), \tag{4}$$

if the following conditions are satisfied

I) The impulsive response of (3), that we call $w^p(t)$, is always positive:

$$(C e^{At} B)^{\frac{1}{p}} = w(t) > 0, \quad 0 \leq t < +\infty \tag{5}$$

II) Called H(t, s) the impulsive response matrix of the error system E (response at time t of the input applied at time s), and $\|H(t, s)\|_{uy}$ its norm induced by the norms $\| \ \|_u$, $\| \ \|_y$ defined in the input and output spaces R_i, R_j, there results

$$\left(\int_{-\infty}^{t} \|H(t, s)\|^q \, w^{-q}(t - s)\, ds \right)^{\frac{1}{q}} = C(t) < \infty, \tag{6}$$

where: $\frac{1}{p} + \frac{1}{q} = 1$ (*).

(*) When $p = 1$, where results $q = \infty$, the condition (6) becomes

$$\max_{-\infty < s \leq t} \|H(t, s)\|_{uy}\, w^{-1}(t - s) = C(t) < \infty$$

The value M(t) = C(t) satisfies the relation (4).

Proof

There holds the following inequality:

$$\|e(t)\|_y = \left\| \int_{-\infty}^{t} H(t, s)\, u(s)\, ds \right\|_y \leqslant \int_{-\infty}^{t} \|H(t, s)\|_{uy} \|u(s)\|_u \, ds \tag{7}$$

Then, because of (5)

$$\|e(t)\|_y \leqslant \int_{-\infty}^{t} \|H(t, s)\|_{uy} \; w^{-1}(t - s)\; w(t - s) \|u(s)\|_u \, ds \tag{8}$$

and considered (1), (6), for the Holder inequality there results

$$\|e(t)\|_y \leqslant \left(\int_{-\infty}^{t} \|H(t, s)\|_{uy}^{q} \; w^{-q}(t-s) ds \right)^{\frac{1}{q}} \left(\int_{-\infty}^{t} w^{p}(t-s) \|u(s)\|_u^{p} ds \right)^{\frac{1}{p}} \tag{9}$$

From (9) the theorem is proved, there follows, in fact:

$$\|e(t)\|_y \leqslant C(t) \quad z'(t) \tag{10}$$

Remark 1

The smallest value $\|E(t)\|$ of M(t) which satisfies (4) is defined by the relation

$$\|E(t)\| = \sup \frac{\|e(t)\|_y}{z'(t)}, \quad u \in U(-\infty, t), \neq 0 \tag{11}$$

Let us, now remark as the output z'(t) of E_m induces a norm into the input function space $U(-\infty, t)$:

$$\|u\|_w = z'(t) = \left(\int_{-\infty}^{t} w^{p}(t - s) (\|u(s)\|_u)^{p} ds \right)^{\frac{1}{p}} \tag{12}$$

With regard to the metric which derives from this norm, the input space $U(-\infty, t)$ defined by (1), does not result complete. The completion of $U(-\infty, t)$ is the space $L^p_w(-\infty, t)$, defined by the property that $z'(t) = \|u\|_w$ results bounded.

Then the relation (11) defines the norm of the linear operator $E(t)$ which model the error system as a mapping from the input space $L^p_w(-\infty, t)$ to the output vector space R_j at the time t,

$$\|E(t)\| = \frac{\|e(t)\|_y}{\|u\|_w}, \quad u \in L^p_w(-\infty, t), \quad \|u\|_w > 0 \tag{13}$$

Of course, there results

$$\|E(t)\| \leq C(t) = \left(\int_{-\infty}^{t} \|H(t, s)\|^q_{uy} \, w^{-q}(t-s) \, ds\right)^{\frac{1}{q}} \tag{14}$$

It is not difficult to verify that in (14) the equality holds always when one input one output systems are considered. Indeed, there results

$$|e(t)| = C(t) \, \|u\|_w$$

when it is:

$$u(s) = \operatorname{sgn}\left[H(t, s)\right] \cdot |H(t, s)|^{\frac{q}{p}} \, w^{-q}(t-s), \quad \text{for } p > 1 \tag{15}$$

$$u(s) = \delta(s - t_o) \qquad \text{for } p = 1$$

where $\delta(t)$ is the impulsive function and t_o the value of s for which the continuous bounded function $|h(t, s)w^{-1}(t-s)|$ has the maximum value.

The equality holds in (14), also for multivariable systems when $p = 1$, but not for $p > 1$. As conclusion, the smallest dynamic upper bound of the error system results

$$\begin{aligned} \dot{x} &= Ax + B\,(\|u(t)\|_u)^p \\ z(t) &= \|E(t)\|\,(C\,x(t))^{\frac{1}{p}} \end{aligned} \tag{16}$$

when A, B, C are given in such way that the system is stable and the conditions (5) (6) are satisfied.

Changing A, B, C we get different upper bounds, which are not comparable among themselves. Indeed, for each one the upper bound is reached by different input signals; that is, the worste case is changing with A, B, C.

Remark 2

The condition (6) can be removed if the following assumption, which involves the set of the acceptable input functions $U(-\infty, +\infty)$, is made

$$\|e'(t)\|_y = \left\| \int_{-\infty}^{t-T} H(t,s)\,u(s)\,ds \right\|_y \leq \delta, \quad u \in U(-\infty, +\infty) \tag{17}$$

where T is given and an error $\|e(t)\|_y \leq \delta$ is disregarded. In fact, following the same procedure applied in the proof of the theorem I, there results:

$$\|e(t)\|_y \leq \|e'(t)\|_y + \left\| \int_{t-T}^{t} H(t,s)u(s)ds \right\|_y \leq \|e'(t)\|_y + \bar{C}(t)\bar{z}'(t) \tag{18}$$

where:

$$\begin{aligned} \bar{C}(t) &= \left(\int_{t-T}^{t} \|H(t,s)\|_{uy}^{q} \; w^{-q}(t-s)\;ds \right)^{\frac{1}{q}} \\ \bar{z}'(t) &= \left(\int_{t-T}^{t} w^{p}(t-s)\;(\|u(s)\|_u)^p\;ds \right)^{\frac{1}{p}} \end{aligned} \tag{19}$$

Then, from (17) there follows:

$$\|e(t)\|_y \leq \delta + \bar{C}(t)\;\bar{z}'(t) \simeq \bar{C}(t)\;\bar{z}'(t) \tag{20}$$

We remark as the condition (6) is no more necessary in order that C(t) exists and is bounded.

The condition (17) which is put on the input functions set, together the assumption that an error $\| e(t) \|_y \leq \delta$ can be disregarded, is equivalent to consider finite the memory of the error system E. From this point of view, the error system norm $\| E(t) \|$, defined in the remark 1 as the smallest value of M(t) which satisfies (4), results:

$$\| E(t) \| = \sup \frac{\| e(t) \|_y}{\| u \|_w}, \quad \text{for } u \in U(t - T,\ t),\ \| u \|_w > 0 \tag{21}$$

Error system norm computing

In the previous section continuous systems are considered: anyway the extension of the results to discrete systems is immediate.

In this section a procedure is developed to evaluate the error system norm following the definition (21), for a discrete linear system described by a set of input-output noisy-free measurements.

For the sake of simplicity a dynamic system S with one input u(i) and one output y(i) is considered.

According to the remark 2 of the previous section, in the following the hypothesis is introduced that the input u(i), for $i < 0$, has no considerable effect in the output at the time $i = n$.

Then, we consider a set of input records $\{u_j(i)\}$ of linear indipendent functions, with $0 \leq i < n$ and $0 < j \leq m$ $(n \geq m)$, and the corresponding outputs $\{y_j(n)\}$ at the time $i = n$.

Chosen a simplified model S' and called y'(i) the model output, the output error for $i = n$ can be computed $\{e_j(n)\} = \{y_j(n) - y'_j(n)\}$. The error system is then, defined by the input records $\{u_j(i)\}$ and by the output vec tor $\{e_j(n)\}$, which are here considered the data of the problem.

The aim is to costruct a dynamic system, the output of which z(n) gives an upper bound of the error, $|e(n)| \leq z(n)$, when the input belongs to the linear manifolds $U[0, n-1]$ spanned by the basis $\{u_j(i)\}$, $0 \leq i < n$, $1 \leq j \leq m$. When m = n the input set considered covers any possible input sequence u(i).

Following the theorem I, we introduce a dynamic system, with one input $u^2(i)$ and one output $z'^2(i)$:

$$x(i + L) = A\, x(i) + B\, u^2(i)$$
$$z'^2(i) = C\, x(i) \qquad (22)$$

where A, B, C are chosen so that the output is always positive and greatest than zero, when the input is different from zero. What can be expressed by the condition

$$C\, A^i\, B = w(i) > 0, \quad i \geq 0$$

In the following the minimum value $\|E(n)\|$ is derived, which satisfies the condition

$$\|E(n)\|\; z'(n) = z(n) \geq |e(n)|, \quad u \in U\,[0, n-1]$$

that is, the error system norm as defined in the previous section.

Let us introduce the following notations

U_{nm} the nxm matrix of the input records $\{u_j(i)\}$

E_m the m-dimension vector of the errors $\{e_j(n)\}$

W_m an nxn diagonal matrix, the elements of which are $w_{ii} = C A^{n-i} B$

α_m an m-dimension real value vector

With the notations introduced, $U_{nm}\alpha_m$ is an input vector belonging to $U\left[0,\ n-1\right]$, $E_m^t\ \alpha_m$ the corresponding error system output at $i = n$, and $(\alpha_m^t\ U_{nm}^t\ W_{nn}\ U_{nm}\ \alpha_m)^{1/2} = z'(n)$ the corresponding output of the system (22). Then the norm of the error system results:

$$\|E(n)\|^2 = \operatorname*{Max}_{\alpha_m} \frac{\alpha_m^t\ E_m\ E_m^t\ \alpha_m}{\alpha_m^t\ U_{nm}^t\ W_{nn}\ U_{nm}\ \alpha_m} \tag{23}$$

or, what is the same:

$$\|E(n)\|^2 = \mathrm{Max}\ (\alpha_m^t\ E_m\ E_m\ \alpha_m),\ \text{for}\ \alpha_m^t\ U_{nm}^t\ W_{nn}\ U_{nm}\ \alpha_m = 1 \tag{24}$$

We have a problem of maximation of a quadratic form with constraints. The application of the Lagrange method leads to the following result:

$$\|E(n)\|^2 = E_m^t\ (U_{nm}^t\ W_{nn}\ U_{nm})^{-1}\ E_m \tag{25}$$

Of course, when S is a time-variant system the norm $\|E(n)\|$ results a function of the instant n considered. Then the computation of (25) have to be repeated changing the final instant considered, with different input records always of lenght n.

On the contrary, for time-invariant systems the norm $\|E(n)\| = \|E\|$ results a constant and it can be evaluated considering input vectors of n samples which are segments of the same input record, also overlapping among themselves in part. If these segments are taken shifted of one sample, the vector E_m results to be the output error vector. An optimisation of the model minimizing the error system norm given by (25), corresponds, then, to apply the minimum square criterion with the weighting matrix $(U_{nm}^t\ W_{nn}\ U_{nm})^{-1}$ which depends from the inputs applied to the system and from the dynamic performance of the system (22).

Conclusion

The uncertainty limits in modeling linear systems by simplified models can be expressed by a dynamic system:

$$\begin{aligned} x(t) &= A\ x(t) + B(\|u(t)\|_u)^p \\ z(t) &= \|E(t)\|\ (C\ x(t))^{1/p} \end{aligned} \tag{26}$$

the output of which z(t) is an upper bound of the error between the model output y'(t) and the actual output y(t),

$$z(t) \geqslant \|y(t) - y'(t)\|_y \tag{27}$$

for any admissible input function.

For any arbitrary choice of the matrix A, B, C, satisfying the conditions stated in section 3, there exists a minimum value $\|E(t)\|$, which is finite, in order that the system (26) has the property (27). Such a value $\|E(t)\|$ is the norm of the error system, defined with regard to a suitable type of norm, depending from the matrix A, B, C, associated to the input function space.

The problem of the best choice of the matrix A, B, C is not well defined, as outlined in the Remark 1, section 3. This problem has not been considered in the note.

The wide freedom of A, B, C seems a good result, since, in this way, the evaluation of the uncertainty limits is reduced to the computation of only one parameter, the norm of the error system $\|E(t)\|$.

The proposed expression of the uncertainty limits may be result very useful in control problem. In fact, when an input function is applied to the actual system, the system (26), associated to the model, permits to compute a tube, centred in the model output and with radius z(t), to which the actual output belongs surely. The tube radius z(t) is the output of the system (26),

Acknowledgments

The author would like to thank M. Milanese for helpful discussions.

This work was supported in part by "Consiglio Nazionale delle Ricerche" and by Nato Grant n. 524.

References

1. F. Donati and M. Milanese: "System identification with approssimated models", 2nd Prague Symposium on Identification and Process Parameter Estimation, Prague, June 1970.
2. F. Donati: "Approssimazione di sistemi lineari in spazi normati" - XI Convegno Automazione e Strumentazione FAST, Milano, nov. 1970.
3. M. Milanese: "Sulla caratterizzazione e identificazione dei sistemi variabili nel tempo", XI Convegno Automazione e Strumentazione FAST, Milano, nov. 1970.
4. M. Milanese: "Identification of uniformly approximating models of system", Ricerche di Automatica, July 1971.
5. M. Milanese, and A. Negro: "Uniform approximation of systems. A Banach space approach" (to appear in JOTA).
6. R. Genesio and R. Pomé: "Spazi funzionali normati e identificazione di sistemi", XI Int. Automation and Instrumentation Conference, FAST, Milano (1970).
7. R. Genesio and R. Pomé: "The approximation of unknown systems by single models. A statistical approach in function spaces" - Preprints of VI International Summer School on Electronics on Automation, Herceg Novi, Yugoslavia, (1971).
8. G. Menga: "Controllo di sistemi complessi approssimati in norma", XI International Automation and Instrumentation Conference, FAST, Milano, (1970).

9. G. Menga and M. Milanese: "Control of system in presence of uncertainty in norm", Preprint of VI International Summer School on Electronics and Automation, Herceg Novi, Yugoslavia, (1971).
10. A. Negro: "Min-max optimal control of systems approximated by finite-dimensional models: non-quadratic cost functional" (to appear in JOTA).

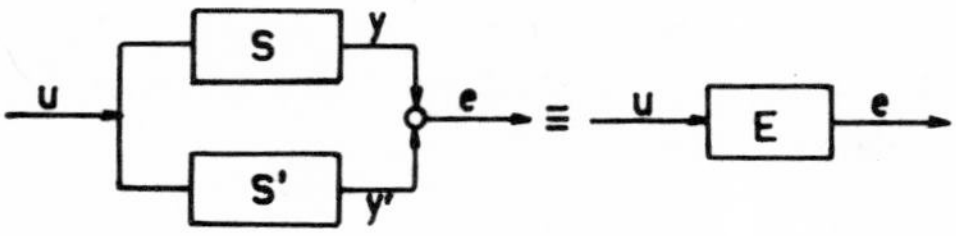

Fig. 1 The error system

SOME RESULTS ON THE ABSTRACT REALIZATION THEORY OF MULTILINEAR SYSTEMS

G. Marchesini
Faculty of Engineering
University of Padova

G. Picci
Faculty of Statistics
University of Padova

Abstract

The state variable description of casual input - output maps of multilinear form is considered. Some structural conditions are indicated under which finite dimensional realization can be obtained. A differential equation description is also given representing the internal behaviour of a bilinear input - output map.

I - Introduction

In many practical problems one is forced to deal with an input-output description fitting the actual behaviour of a physical process. This is a typical situation arising in cases where the complexity of the internal structure prohibits any derivation of a mathematic model from first principles.

The existing mathematical techniques constrain most cases to adopt linear input - output descriptions. This is due to the lack of efficient identification methods and in general to the very poor understanding of the behaviour of systems described by nonlinear input-output (i. e functional) models.

Despite this, if one is convinced that a linear input-output description does not satisfy the needs of the analysis of the physical process, one could try to add non-linear terms in order to obtain a better fit to the actual behaviour.

A typical situation is described by a relationship of the following form:

$$y(t) = \int_{t_o}^{t} k_1(t,\tau)\, u(\tau)\, d\tau + \dots \tag{1.1}$$

$$\dots + \int_{t_o}^{t} \dots \int_{t_o}^{t} k_n(t,\tau_1 \dots \tau_n)\, u(\tau_1) \dots u(\tau_n)\, d\tau_1 \dots d\tau_n,$$

where the terms like $\int_{t_o}^{t} \dots \int_{t_o}^{t} k_i(t,\tau_1 \dots \tau_i)\, u(\tau_1) \dots u(\tau_i)\, d\tau_1 \dots d\tau_i$

This work was supported by CNR. Automatic Control Group.

play the role of nonlinear correctors.

Although some efforts have been made to investigate such input-output relationship one cannot find in the literature any feasible technique either for analyzing systems of this structure in closed feedback loops or for developing efficient closed loop control strategies. In addition a serious practical drawback exists, i.e. the well known computational difficulty associated with high order convolution type representations.

The previous observations justify some efforts devoted to "state variabilizing" the input-output functional polynomial model (1.1). This corresponds substantially to trying to find explicit conditions for a differential equation representation, and at the same time, to obtain corresponding computational algorithms.

The problem has been investigated in the past by Balakrishnan [1].

M. A. Arbib [2] attacked virtually the same problem with an entirely different approach. He obtained a decomposition for multilinear discrete time constant systems in terms of linear subsystems and multipliers. A similar representation was heuristically used in the past by Schetzen [3] and Bush [4].

In this paper the idea indicated by Arbib is further developed to obtain explicit conditions for minimal realizations of time varying multilinear maps.

2 - Statement of the problem

This paper will be concerned only with the second order term in (1.1). We shall do this mainly in order to deal with tractable formalism. The underlying line of reasoning will be easily recognized as being valid for higher order terms too.

We shall consider then the second order term

$$\int_{t_o}^{t}\int_{t_o}^{t} k(t,\tau,\sigma)\; u(\tau)\; u(\sigma)\; d\tau\; d\sigma \tag{2.1}$$

as the image of an appropriate family of input-output maps

$$f_t \;:\; U_t \longrightarrow R^{I} \qquad t \in T \tag{2.2}$$

acting from an input space U into an output space Y as the following definitions puntualize:

a) the time set T

The time set is an inferiorly bounded interval of the real line:

$$T = \{t \geqslant t_o \;\;,\;\; t_o > -\infty\} \tag{2.3}$$

b) the input space U

The input space U is the set of real valued piecewise continuous functions defined over T. The main feature of the input space which is of interest here, is the algebraic R - vector space structure under the usual operations.

In (2.2) a particular subspace $U_t \subset U$ is considered:

$$U_t = \{u \in U : u(\tau) = o \quad \forall \tau \geq t \; ; \; \tau, \; t \in T\} \qquad (2.4)$$

furthermore, if $U_{t_i t_{i+1}}$ denotes the subspace :

$$U_{t_i t_{i+1}} = \{u \in U : u(\tau) = o \; \forall \tau < t_i, \; \forall \tau \geq t_{i+1}; \; t_{i+1} \geq t_i \in T\} \qquad (2.5)$$

then the direct sum decomposition:

$$U_t = U_{t_o t_1} \oplus U_{t_1 t_2} \oplus \; \ldots . \; \oplus U_{t_{n-1} t_n} \qquad (2.6)$$

holds, where $t_o \leq t_1 \leq \ldots t_{n-1} \leq t_n = t$.

The above decomposition always makes possible the concatenation of inputs by means of the sum of elements belonging to suitable subspaces $U_{t_i t_j}$.

c) the ouput space Y

The space Y is assumed to be the R - vector space of real valued functions continuous over T.

A particular subspace Y_t of Y is also defined as

$$Y_t = \{y \in Y : y(\tau) = o, \; t_o \leq \tau < t\} \qquad (2.7)$$

d) the causal kernel k

The kernel function k :

$$k : (T \times T^2) \longrightarrow R^1 \qquad (2.8)$$

is continuous at every point of the set $(T \times T^2)^c$, defined by :

$$(T \times T^2)^c : \{(t, \tau, \sigma) : t \geq \max(\tau, \sigma), \; t \in T, \; (\tau, \sigma) \in T^2\} \qquad (2.9)$$

and zero outside.

Under the above assumptions the map (2.2) will be referred to as a causal input - output quadratic map. Note that it can be viewed as a second degree homogeneous polynomial in u.

This quadratic map can be biuniquely associated with a **bilinear** map defined by the same kernel k. This map acts from $U_t \times U_t$ into R^1 via the relationship:

$$y(t) = \int_{t_o}^{t} \int_{t_o}^{t} k(t, \tau, \sigma) \, u(\tau) \, v(\sigma) \, d\tau \, d\sigma \qquad t \geq t_o \qquad (2.10)$$

$$(u \, , \, v) \in U_t x U_t$$

The relation (2.10) is rewritten in a more compact form as :

$$y(t) = f_t \binom{u}{v} \; ; \qquad f_t : U_t \times U_t \longrightarrow R^1 \qquad (2.11)$$

which reflects a natural interpretation of u as the input in the "first input channel" and v as the input in the "second input channel" of a system described by the input-output bilinear relationship (2.10).

The problem is now stated in the following way:

PROBLEM. *(Realization) Find, if it exists, a dynamical system in state space form such that it generates the same input-output pairs as the given input-output map* (2.10).

The solution of this problem is developed along the following lines:

1) Introduce the notion of the "state of the system" at each instant $s \in T$.
2) Construct the state set, X_s, at the instant s.
3) Find a state transition map, F

$$F(t,s,.,.,.) : X_s \times U_{st} \times U_{st} \longrightarrow X_t \qquad t > s. \tag{2.12}$$

4) Find a read-out map, G,

$$G(t,.) : X_t \longrightarrow R^1 \tag{2.13}$$

5) Define a zero state at the instant t_o, x_o, such that

$$G[t,F(t,t_o,x_o,0,0)] = 0 \qquad t > t_o \tag{2.14}$$

6) X_s, F, G and x_o have to be found such that the ouput values:

$$y(t) = G\,[t,F(t,t_o,\ x_o,\ u,v)] \qquad t > t_o \tag{2.15}$$

are exactly the same as those generated by (2.10).

In other words X_s, F, G and x_o make the following diagram

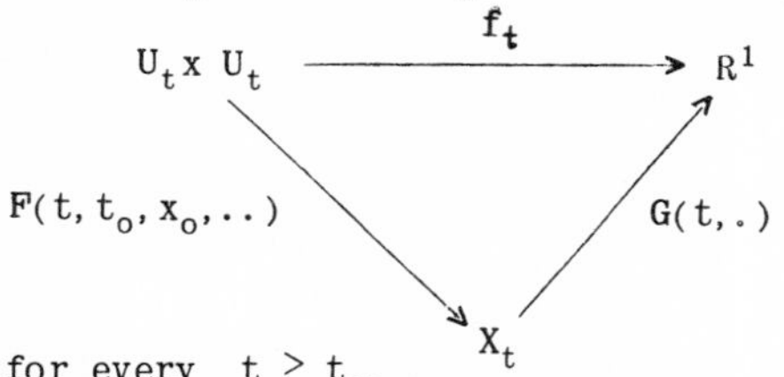

(2.16)

commutative, for every $t > t_o$.

It is known, [5.6], that the realization problem can be trivially solved taking the input space, $U_s \times U_s$, as the state space X_s. Of course this state space model is unduly complex since it requires to store the complete past inputs. So, in order to obtain efficient realizations it is worthwhile looking for structures for which the state space **X** has as few elements as possible.

Firstly, no equivalent states must exist: starting from different states in X_s different responses in Y_s have to be obtained. More precisely, for each pair $x_s \neq x'_s$ of distinct states in X_s there is at least one input pair $(z,w) \in U_{st} \times U_{st}$ to wich they react differently. Such a state space is said to be **completely observable.**

Secondly, every state in X_s must come from some input pair in $U_s \times U_s$ (once fixed the zero state x_o at t_o). This fact implies that, in the diagram (2.16), F is a surjective map and, in this situation, the state space is referred to as **completely reachable.**

It is relevant now the following

DEFINITION. *A realization is said to be canonical if the state space is both completely reachable and observable.*

A general approach to the problem of finding a canonical realization of the input-output map f_t is supplied by the fundamental Theorem of Nerode for automaton minimization [7], which in the continuous time case is rephrased as:

THEOREM. *Given* $f_t : U_t x U_t \longrightarrow R^1$, *define over* $U_s x\, U_s$, $s \leqslant t$, *the equivalence relation* $\overset{s}{\sim}$ *by*:

$$\begin{pmatrix} u \\ v \end{pmatrix} \overset{s}{\sim} \begin{pmatrix} u' \\ v' \end{pmatrix} \quad \text{iff} \quad f_t \begin{pmatrix} u & z \\ v & w \end{pmatrix} = f_t \begin{pmatrix} u' & z \\ v' & w \end{pmatrix} \tag{2.17}$$

$\forall\, t \geqslant s$, $\forall\, (z,w) \in U_{st} x\, U_{st}$,

and the equivalence class $\begin{bmatrix} u \\ v \end{bmatrix}(s)$ *by:*

$$\begin{bmatrix} u \\ v \end{bmatrix}(s) = \left\{ \begin{pmatrix} u' \\ v' \end{pmatrix} \in U_s \times U_s \;:\; \begin{pmatrix} u' \\ v' \end{pmatrix} \overset{s}{\sim} \begin{pmatrix} u \\ v \end{pmatrix} \right\} \tag{2.18}$$

The quotient space associated with the $\overset{s}{\sim}$ *equivalence is then given by*

$$X_s = (U_s x\, U_s) \,/ \overset{s}{\sim} = \left\{ \begin{bmatrix} u \\ v \end{bmatrix}(s) \;:\; \begin{pmatrix} u \\ v \end{pmatrix} \varepsilon U_s x\, U_s \right\} \tag{2.19}$$

Let, now, the dynamical system $S(f) : (T, U_{st} x\, U_{st}, X_s, Y, F, G)$ *be defined by the following relationship:*

i) *The state transition map* $F(t,s,.,.,.,.) : X_s x\, U_{st} x\, U_{st} \longrightarrow X_t$

$$F(t,s,.,.,.,.) : \left(\begin{bmatrix} u \\ v \end{bmatrix}(s), \begin{pmatrix} z \\ w \end{pmatrix} \right) \mapsto \begin{bmatrix} u & z \\ v & w \end{bmatrix}(t) \tag{2.20}$$

ii) *The readout map* $G : X_t \longrightarrow R^1$

$$G(t,.) : \begin{bmatrix} u & z \\ v & w \end{bmatrix}(t) \longmapsto f_t \begin{pmatrix} u & z \\ v & w \end{pmatrix} \tag{2.21}$$

Then the dynamical system $S(f)$ *is a canonical realization of the input-output map* f_t.

REMARKS. Althought the Theorem clarifies how the realization problem must be set out in an appropriate context, many questions remain to be answered. The principal ones are:

1 - what is the algebraic structure of the state space X_s? Can X_s be endowed with a vector space structure? If the answer is negative, does there exist some geometric way to characterize X_s?

2 - if X_s exhibits a vector space structure, is it finite dimensional? If the answer is positive is it possible under suitable smoothness

assumptions (see for instance [5]) to describe the system S(f) by a differential equation?

3 - If X_s is a finite dimensional vector space how can one derive a corresponding differential equation description? Could the answer to this question be an algorithm allowing the evaluation of the differential equation model starting from $k(t,\tau,\sigma)$?

Some efforts have been made by M.A. Arbib [2] to answer these questions in the area of discrete time constant systems. We shall attempt to proceed further by dealing mainly with the problems indicated by the questions 2 and 3 for the continuous time-varying system (2.10).

3 - Decomposition

This section is devoted to the drawing out of some important consequences of the bilinear (or more generally multilinear) structure of f_t. The main goal will be the tranfer of the R-vector space structure of the input space onto the set of states as defined in the Nerode realization theory.

As a final step in this direction, X_s is decomposed into a triple (X_{1s}, X_{2s}, X_{3s}) of sets, whose elements still qualify as states. This fact is essentialy a consequence of the following decomposition property of f_t :

$$\int_{t_o}^{t}\int_{t_o}^{t} k(t,\tau,\sigma)u^*(\tau)v^*(\sigma)d\tau d\sigma = \int_{t_o}^{s}\int_{t_o}^{s} k(t,\tau,\sigma)u(\tau)v(\sigma)\, d\tau\, d\sigma + \int_{t_o}^{s}\int_{s}^{t} k(t,\tau,\sigma)u(\tau)\, w(\sigma)\, d\tau d\sigma + \int_{s}^{t}\int_{t_o}^{s} k(t,\tau,\sigma)z(\tau)v(\sigma)d\tau d\sigma + \int_{s}^{t}\int_{s}^{t} k(t,\tau,\sigma)z(\tau)w(\sigma)d\tau d\sigma \tag{3.1}$$

where :

$$u^* = u + z \qquad v^* = v + w \qquad \begin{array}{lll} u^*, v^* & \varepsilon & U_t \\ u,\ v & \varepsilon & U_s \\ z,\ w & \varepsilon & U_{st} \end{array} \tag{3.2}$$

It will be helpful for understanding what follows to examine in some detail the structure of the four terms appearing in the R.H.S. of (3.1).

The output y(t) appears to be sum of four contributions:

i) a "free" response (no inputs are applied after the instant s) starting from the situation into which the system was carried by the inputs $(u,v)\ \varepsilon\ U_s \times U_s$, acting up to s.

ii) the response of the system as driven up to time s only by the first channel input u, and then, after s, only by the second channel input $w\ \varepsilon\ U_{st}$. Clearly at the instant s, the system moves from the "state" into which it was driven by the input u.

iii) a response corresponding to the exact symmetric situation as in the case ii).

iv) a "forced" response which is zero before the time because of the causality of the kernel k. Since zero inputs are applied before s, the response starts from a "quiescent" situation.

We rewrite now (3.1) to evidentiate the bilinearity of f_t :

$$f_t\begin{pmatrix}u^*\\v^*\end{pmatrix} = f_t\begin{pmatrix}u & o\\v & o\end{pmatrix} + f_t\begin{pmatrix}u & o\\o & w\end{pmatrix} + f_t\begin{pmatrix}o & z\\v & o\end{pmatrix} + f_t\begin{pmatrix}o & z\\o & w\end{pmatrix} \tag{3.3}$$

The relationship (3.3) suggests the following equivalences:

$$u \overset{s}{\sim}_1 u' \quad \text{iff} \quad f_t\begin{pmatrix}u & o\\o & \varphi\end{pmatrix} = f_t\begin{pmatrix}u' & o\\o & \varphi\end{pmatrix} \qquad \forall t \geq s,\ \forall \varphi \in U_{st} \tag{3.4}$$

$$v \overset{s}{\sim}_2 v' \quad \text{iff} \quad f_t\begin{pmatrix}o & \theta\\v & o\end{pmatrix} = f_t\begin{pmatrix}o & \theta\\v' & o\end{pmatrix} \qquad \forall t \geq s,\ \forall \theta \in U_{st} \tag{3.5}$$

$$\begin{pmatrix}u\\v\end{pmatrix} \overset{s}{\sim}_3 \begin{pmatrix}u'\\v'\end{pmatrix} \quad \text{iff} \quad f_t\begin{pmatrix}u & o\\v & o\end{pmatrix} = f_t\begin{pmatrix}u' & o\\v' & o\end{pmatrix} \qquad \forall t \geq s \tag{3.6}$$

$u, u' \in U_s$, $v, v' \in U_s$.

by which the equivalence classes are introduced:

$$[u]_1(s) = \{u' : u' \overset{s}{\sim}_1 u\} \tag{3.7}$$

$$[v]_2(s) = \{v' : v' \overset{s}{\sim}_2 v\} \tag{3.8}$$

$$\begin{bmatrix}u\\v\end{bmatrix}_3 (s) = \left\{\begin{pmatrix}u'\\v'\end{pmatrix} : \begin{pmatrix}u'\\v'\end{pmatrix} \overset{s}{\sim}_3 \begin{pmatrix}u\\v\end{pmatrix}\right\} \tag{3.9}$$

It is easy to prove that Nerode equivalence implies and is implied by the previous three equivalences but we shall not pursue this point further.

Define now the following quotient spaces:

$$X_{1s} = U_s/\overset{s}{\sim}_1 = \{[u]_1(s) : u \in U_s\} \tag{3.10}$$

$$X_{2s} = U_s/\overset{s}{\sim}_2 = \{[v]_2(s) : v \in U_s\} \tag{3.11}$$

$$X_{3s} = (U_s \times U_s)/\overset{s}{\sim}_3 = \left\{\begin{bmatrix}u\\v\end{bmatrix}_3 (s) : \begin{pmatrix}u\\v\end{pmatrix} \in U_s \times U_s\right\} \tag{3.12}$$

which in our decomposition will play the role of state spaces.

We start with the statement of the

LEMMA 1. *The equivalence relations* $\overset{s}{\sim}_1$, $\overset{s}{\sim}_2$ *are congruences and*

$$u \overset{s}{\sim}_1 u' \Longrightarrow u + z \overset{t_1}{\sim}_1 u' + z \qquad \forall z \in U_{st_1}, \forall t_1 \geq s \tag{3.13}$$

$$v \overset{s}{\sim}_2 v' \Longrightarrow v + w \overset{t_1}{\sim}_2 v' + w \qquad \forall w \in U_{st_1}, \forall t_1 \geq s \tag{3.14}$$

proof. Only the proof of (3.13) is given. The proof of (3.14) is the same. Firstly recall that:

$$u \overset{s}{\sim}_1 u' \text{ iff } f_t \begin{pmatrix} u & o \\ o & \varphi \end{pmatrix} = f_t \begin{pmatrix} u' & o \\ o & \varphi \end{pmatrix} \qquad \begin{matrix} \forall \varphi \in U_{st} \\ \forall t \geq s \end{matrix} \tag{3.15}$$

Let us consider the subspace $U_{t_1 t}$ of U_{st} corresponding to some fixed $t_1 \geq s$. If ψ is any function belonging to $U_{t_1 t}$ then (3.15) implies:

$$f_t \begin{pmatrix} u & o & o \\ o & o & \psi \end{pmatrix} = f_t \begin{pmatrix} u' & o & o \\ o & o & \psi \end{pmatrix} \qquad \begin{matrix} \forall \psi \in U_{t_1 t} \\ \forall t \geq t_1 \end{matrix} \tag{3.16}$$

so that $uo \overset{t_1}{\sim}_1 u'o$.

Adding the quantity

$$f_t \begin{pmatrix} o & z & o \\ o & o & \psi \end{pmatrix} \qquad z \in U_{st_1} \tag{3.17}$$

to both terms in (3.16) one obtains:

$$f_t \begin{pmatrix} u & z & o \\ o & o & \psi \end{pmatrix} = f_t \begin{pmatrix} u' & z & o \\ o & o & \psi \end{pmatrix} \qquad \begin{matrix} \forall \psi \in U_{t_1 t} \\ \forall t \geq t_1 \end{matrix} \tag{3.18}$$

which holds for any z belonging to U_{st_1}. This proves the Lemma.

OBSERVATION. *A direct consequence of the previous result is :*

$$\text{if} \quad z \overset{t_1}{\sim}_1 z' \qquad \text{then} \quad u + z \overset{t_1}{\sim}_1 u' + z' . \tag{3.19}$$

The fact that the equivalence relations $\overset{s}{\sim}_1$ and $\overset{s}{\sim}_2$ are congruences allows to transport onto the quotient spaces X_{1s} and X_{2s} the algebraic structure of U_s. That is, X_{1s} and X_{2s} can be made vector spaces under the operations:

$$[u']_1(s) + [u'']_1(s) \triangleq [u' + u'']_1(s) \qquad u', u'' \in U_s \tag{3.20}$$

$$\alpha\, [u]_1(s) \triangleq [\alpha u]_1(s) \qquad \alpha \in R^1$$

and

$$[v']_2(s) + [v'']_2(s) \triangleq [v' + v'']_2(s) \qquad v', v'' \in U_s$$
$$\alpha[v]_2(s) \triangleq [\alpha v]_2(s) \qquad \alpha \in R^1 \tag{3.21}$$

where the symbols used in the right hand sides make sense since they are independent on the particular choice of the elements in the equivalence classes.

The vector space structure of X_{1s} and X_{2s} gives immediately:

$$[uz]_1(t_1) = [uo + oz]_1(t_1) = [uo]_1(t_1) + [oz]_1(t_1) \tag{3.22}$$

$$[vw]_2(t_1) = [vo + ow]_2(t_1) = [vo]_2(t_1) + [ow]_2(t_1) \tag{3.23}$$

The significance of relationships (3.22), (3.23), is now drawn out by means of the

LEMMA 2. *Define the two maps:*

$$F_1(t_1, s) : X_{1s} \longrightarrow X_{1t_1} \tag{3.24}$$
$$t_1 \geqslant s$$
$$F_2(t_1, s) : X_{2s} \longrightarrow X_{2t_1} \tag{3.25}$$

in such way that :

$$[uo]_1(t_1) = F_1(t_1, s)[u]_1(s) \tag{3.26}$$
$$t_1 \geqslant s$$
$$[vo]_2(t_1) = F_2(t_1, s)[v]_2(s) \tag{3.27}$$

then F_i, $i=1,2$, *are well defined, linear and enjoy the property:*

$$F_i(t_2, s) = F_i(t_2, t_1) \circ F_i(t_1, s) \qquad t_2 \geqslant t_1 \geqslant s, \quad i = 1,2 \tag{3.28}$$

proof. The linearity is an immediate consequence of Lemma 1.

In fact suppose that $[u]_1(s) = [u']_1(s) + [u'']_1(s)$ then by (3.13):

$$[uo]_1(t_1) = [(u'+u'')o]_1(t_1) = [u'o+u''o]_1(t_1) = [u'o]_1(t_1) + [u''o]_1(t_1). \tag{3.29}$$

The property (3.28) follows from:

$$\begin{aligned}[uzo]_1(t_2) &= F_1(t_2, t_1) \circ F_1(t_1, s)[u]_1(s) + F_1(t_2, t_1)[oz]_1(t_1) \\ &= F_1(t_2, s)[u]_1(s) + [ozo]_1(t_2) = \\ &= F_1(t_2 s)[u]_1(s) + F_1(t_2 t_1)[oz]_1(t_1) \qquad \forall\, t_2 \geqslant t_1 \geqslant s.\end{aligned} \tag{3.30}$$

which holds for any $u \in U_s$, $z \in U_{st_1}$.

Doing so two linear systems in the state space form have been extracted from the input-output map f_t.

Let us consider now the quotient space X_{3s} obtained modulo the equivalence $\overset{s}{\sim}_3$:

$$X_{3s} = (U_s \times U_s) \,/\, \overset{s}{\sim}_3 = \left\{ \begin{bmatrix} u \\ v \end{bmatrix}_3 (s) \;:\; \begin{pmatrix} u \\ v \end{pmatrix} \in U_s \times U_s \right\} \tag{3.31}$$

OBSERVATION. The space X_{3s} does not admit a vector space structure with operations like (3. 20). This is a direct consequence of the nonlinearity of f_t as acting contemporaneously on both input channels.

More precisely, if we define the map

$$f_s : U_s \times U_s \longrightarrow Y_s \tag{3.32}$$

as

$$f_s \begin{pmatrix} u \\ v \end{pmatrix} = \left\{ f_t \begin{pmatrix} u \\ v \end{pmatrix}, \quad t \geqslant s \right\} \tag{3.33}$$

one easily recognizes that the quotient set X_{3s} can be defined through f_s by using the equivalence induced by f_s over $U_s x U_s$. So it is the nonlinearity of f_s which causes the drawback.

The map f_s may be considered to be linear if the input space $U_s \times U_s$ is embedded in a larger space. This is a typical situation in the theory of multilinear products of abstract algebra [8]. To make the procedure that will be adopted to enlarge the input space clearer, it seems convenient to recall here the basic proposition:

PROPOSITION. *Let* U, V *be fixed* R-*vector spaces, then there exist a bilinear map* $\otimes$ *(a tensor product) and a* R-*vector space* $U \otimes V$ *such that for any bilinear map* $f : U \times V \longrightarrow Y$, *there exists a linear map* $f^* : U \otimes V \longrightarrow Y$ *such that the following diagram*

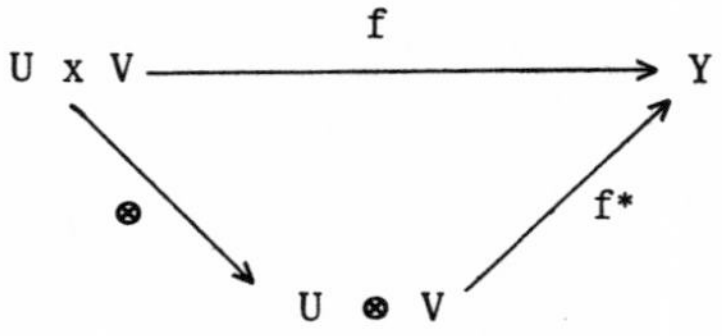

(3.34)

is commutative.

We attempt now to exploit this result to our purposes. To do this define the tensor product in the following way.

For any $\binom{u}{v} \in U_s \times U_s$ define the two variables function:

$$\mu \quad : [t_o s) \times [t_o s) \longrightarrow R^1 \tag{3.35}$$

$$\mu\,(\tau,\sigma) = u(\tau)\cdot v(\sigma) \qquad \forall\, (\tau,\sigma) \in [t_o s) \times [t_o s)$$

Introduce now the vector space U^*_{ss} generated by the set M_s of all μ functions as:

$$U^*_{ss} = \left\{ \varphi(.,.) : \varphi(.,.) = \sum_{1}^{n} \alpha_i \mu_i (.,.), \forall \alpha_i \in R^1, \mu_i \in M_s \right\} \quad (3.36)$$

We recognize immediately that $U^*_{ss} = U_s \otimes U_s$. Beyond that, $f^*_s : U^*_{ss} \longrightarrow Y_s$ is the linear map whose restriction to M_s is

$$f^*_s | M_s : \mu \longmapsto f_s \binom{u}{v} \quad (3.37)$$

where u and v are such that $\mu(.,.) = u(.).v(.)$.

The linear map f^*_s is written explicitely as:

$$f^*_s(\varphi) = \int_{t_o}^{s} \int_{t_o}^{s} k(.,\tau,\sigma)\, \varphi(\tau,\varphi)\, d\tau d\sigma, \qquad \varphi \varepsilon U^*_{ss} \quad (3.38)$$

As we are dealing with tensor products we refer here also the following fact, which shall be useful in the sequel: the direct sum decomposition of the input space (2.6) gives ([8] pag. 413)

$$U_{t_1} \otimes U_{t_1} = (U_s \otimes U_s) \oplus (U_s \otimes U_{st_1}) \oplus (U_{st_1} \otimes U_s) \oplus (U_{st_1} \otimes U_{st_1}) \quad t_1 \geqslant s, \quad (3.39)$$

or, more compactly :

$$U^*_{t_1 t_1} = U^*_{ss} \oplus U^*_{s,st_1} \oplus U^*_{st_1,s} \oplus U^*_{st_1,st_1} \quad (3.40)$$

where the linear subspaces:

$$U^*_{s,st_1} = \left\{ \varphi \varepsilon U_{t_1} \otimes U_{t_1} : \varphi(\tau,\sigma) = 0, \quad \tau > s, \sigma \notin [st_1) \right\}$$

$$U^*_{st_1,s} = \left\{ \varphi \varepsilon U_{t_1} \otimes U_{t_1} : \varphi(\tau,\sigma) = 0, \quad \tau \notin [st_1), \sigma > s \right\} \quad (3.41)$$

$$U^*_{st_1,st_1} = \left\{ \varphi \varepsilon U_{t_1} \otimes U_{t_1} : \varphi(\tau,\sigma) = 0, \quad (\tau,\sigma) \notin [st_1) \times [st_1) \right\}$$

have been introduced.

The previous statements allow us to define a congruence relation on U^*_{ss} as:

$$\varphi \overset{*s}{\sim}_3 \varphi' \qquad \text{iff} \quad f^*_s(\varphi) = f^*_s(\varphi') \quad (3.42)$$

The corresponding quotient space $X^*_{3s} = U^*_{ss} / \overset{*s}{\sim}_3$ is then:

$$X^*_{3s} = \left\{ [\varphi]_3(s) : \varphi \varepsilon U^*_{ss} \right\} \quad (3.43)$$

The linearity of f^*_s allows the space X^*_{3s} to be made a R- vector space with the usual operations:

$$
\begin{aligned}
[\varphi']_3(s) + [\varphi'']_3(s) &\triangleq [\varphi' + \varphi'']_3(s) \\
\alpha\,[\varphi]_3(s) &\triangleq [\alpha\varphi]_3(s)
\end{aligned}
\tag{3.44}
$$

As a consequence, we are naturally allowed to apply the same arguments which originated Lemma 1, having a mind to find updating formulas for the equivalence classes. To this end we recall that the congruence relation $\overset{*t_1}{\sim}_3$, for any $t_1 \geqslant s$, defines a canonical homomorphism π_{t_1} from $U^*_{t_1t_1}$ onto $X^*_{3t_1}$ which maps each element $\varphi \in U^*_{t_1t_1}$ into the equivalence class $[\varphi]_3(t_1)$. Furthermore the direct sum decomposition (3.40) of $U^*_{t_1t_1}$ allows the canonical homomorphism π_{t_1} to be split via the restrictions on the corresponding subsoaces:

$$
\begin{aligned}
U^*_{ss} &\xrightarrow{\;\pi'_{t_1}\;} X^*_{3t_1} \\
U^*_{s,st_1} &\xrightarrow{\;\pi''_{t_1}\;} X^*_{3t_1} \\
U^*_{st_1,s} &\xrightarrow{\;\pi'''_{t_1}\;} X^*_{3t_1} \\
U^*_{st_1,st_1} &\xrightarrow{\;\pi^{IV}_{t_1}\;} X^*_{st_1}
\end{aligned}
\tag{3.45}
$$

At this point we may refer to the linearity of π_{t_1} on $U^*_{t_1,t_1}$ and write

$$
\pi_{t_1}(\varphi) = \pi'_{t_1}(\varphi_{ss}) + \pi''_{t_1}(\varphi_{s\,st_1}) + \pi'''_{t_1}(\varphi_{st_1s}) + \pi^{IV}_{t_1}(\varphi_{st_1,st_1})
\tag{3.46}
$$

where φ_{ss}, φ_{s,st_1}, $\varphi_{st_1,s}$ and φ_{st_1,st_1} are the (unique) projections of φ on U^*_{ss}, U^*_{s,st_1}, $U^*_{st_1,s}$, $U^*_{st_1,st_1}$ respectively.

The updating relationship on the state space $X^*_{3t_1}$ is derived factorizing the map π'_{t_1} via the preceding state space X^*_{3s}, [5] :

$$
\begin{array}{ccc}
U^*_{ss} & \xrightarrow{\;\pi'_{t_1}\;} & X^*_{3t_1} \\
 & {\scriptstyle \pi'_s}\searrow \qquad \nearrow {\scriptstyle F_3(t_1,s)} & \\
 & U^*_{ss}/\overset{*s}{\sim}_3 &
\end{array}
$$

A suitable factorization is obtained by choosing π'_s as the canonical homomorphism induced by the congruence $\overset{*s}{\sim}_3$ and the map $F_3(t_1, s)$ as in the following

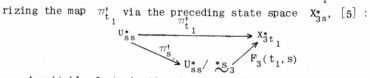

LEMMA 3. *There exists a linear map*

$$F_3\ (t_1, s)\ :\ X^*_{3s} \longrightarrow X^*_{3t_1} \tag{3.48}$$

such that;

$$[\varphi 0]_3(t_1) = F_3(t_1, s)\ [\varphi]_3\ (s) \qquad \varphi \in U^*_{ss} \tag{3.49}$$

and

$$F_3\ (t_2, s) = F_3\ (t_2, t_1) \circ F_3\ (t_1, s) \qquad \forall\ t_2 \geqslant t_1 \geqslant s. \tag{3.50}$$

proof. The proof proceeds as in Lemma 2.

The conclusion to draw from the preceding results may be phrased as follows:

PROPOSITION. *Let* $x_3\ (t_1)$ *be any element of* $X^*_{3t_1}$. *Then there exist*

$$\varphi_{s, st_1} \in U_s \otimes U_{st_1},\ \varphi_{st_1, s} \in U_{st_1} \otimes U_s,\ \varphi_{st_1, st_1} \in U_{st_1} \otimes U_{st_1}$$

such that:

$$x_3(t_1) = F_3(t_1 s)\ x_3(s) + \pi''_{t_1}\ \varphi_{s, st_1} + \pi'''_{t_1}\ \varphi_{st_1, s} + \pi^{IV}_{t_1}\ \varphi_{st_1, st_1} \tag{3.51}$$

REMARK. The relation (3.51) has been obtained without explicit reference to actual inputs. It is worthwhile examining now how the inputs enter the updating relation derived above. To do that, one has to take into account only the subset of U^*_{ss} constituted by the image M_s of the bilinear map $\otimes$.

This situation is clarified by the diagram

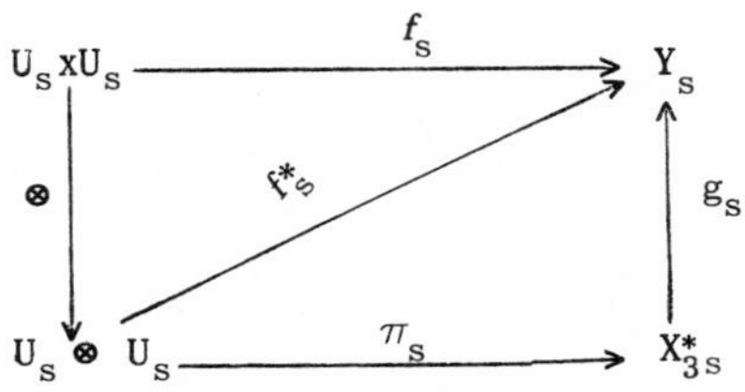

(3.52)

where the linear map f^*_s has been factorized through the quotient space X^*_{3s} .

We recall that π_s is the canonical homomorphism from U^*_{ss} onto X^*_{3s} and g_s is the injection defined by

$$g_s\ :\ [\varphi]_3(s) \longmapsto f^*_s\ (\varphi')\ ,\quad \varphi' \in [\varphi]_3(s) \tag{3.53}$$

Let us imagine f^*_s as the input-output map of a system acting from input space U^*_{ss} into Y_s. Then the canonical factorization $f^*_s = g_s \circ \pi_s$ of f^*_s, may be viewed as a canonical (i.e. completely reachable and observable) realization. This agrees with the meaning given to diagram (2.16). Actually the space U^*_{ss} is not the input space for f_s and only a subset, $M_s \subset U^*_{ss}$, is "reachable" from $U_s \times U_s$. This set is the range of the bilinear map $\otimes$, so if we insist on taking X^*_{3s} as the state space only the states belonging to $\pi_s(M_s)$ constitute the reachable subset of X^*_{3s}.

This kind of arguments makes clear that the realization of f_s pictured by diagram (3.52) is not canonical. We could then ask ourselves if a better choice for the state space could have been made. A partial answer for this question is supplied by the following

PROPOSITION: X^*_{3s} *is the smallest* R-*Vector space containing the reachable set* $\pi_s(M_s)$.

Proof. The proof follows immediately once recalled (3.36) and (3.44).

We attempt now to investigate briefly if we can forego the vector space structure and refer to the reachable manifold $\pi_s(M_s) \subset X^*_{3s}$, as a canonical state space.

With a view to deriving the updating equations into $\pi_s(M_s)$, it is easy to convince ourselves that it would be necessary to know the geometric structure of $\pi_s(M_s)$ itself. From this point of view only the choice of a suitable curvilinear coordinate system into $\pi_s(M_s)$, would permit to obtain the state equation in explicit form. In any case this would not ensure the linearity of the realization.

This is the reason why we shall constrain ourselves to X^*_{3s}, in order to obtain linear structures. Obviously the realization we shall obtain will not be canonical.

4. A finite dimensional realization

We come back, now, to consider the specific functional structure of the bilinear map in order to make our decomposition results more specific. The direction along which we intend now to proceed is that leading to finite-dimensionality conditions for the state space. As we shall see this corresponds directly to a differential equation description for the bilinear map (2.10).

First of all we transform the state space through an isomorphism.

Let Σ_s be any subspace of U^*_{ss} supplement to $\{\ker f^*_s\}$ so that:

$$U^*_{ss} = \ker f^*_s \otimes \Sigma_s \tag{4.1}$$

then there exists the surjective homomorphism (projection map)

$$P_s : U^*_{ss} \longrightarrow \Sigma_s \quad ; \quad P_s(U^*_{ss}) = \Sigma_s \tag{4.2}$$

Introducing the map $H_s : \Sigma_s \longrightarrow Y_s$ defined by:

$$H_s = f^*_s \mid \Sigma_s \tag{4.3}$$

we have a canonical factorization of f^*_s through Σ_s represented by the commutative diagram

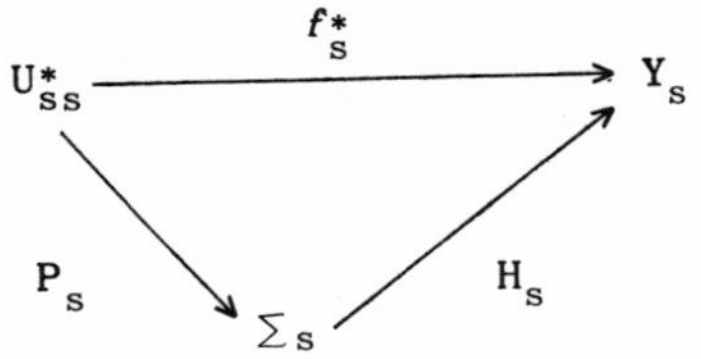

(4.4)

Recalling (3.52), by a well known reasoning (see for instance [5] pag. 258) we have the following

PROPOSITION. *The linear vector spaces* Σ_s *and* X^*_{3s} *are isomorphic.*

In view of this fact from now on we shall assume Σ_s as the state space.

The conditions for finite dimensionality of Σ_s are now phrased by the following.

LEMMA 4. *The state space* Σ_s *is isomorphic to* $\{\mathrm{Im}\ f^*_s\}$. *As a consequence it is finite dimensional iff the map* $f^*_s : U^*_{ss} \longrightarrow Y_s$, *where*

$$f^*_s\ (\varphi) = \int_{t_o}^{s} \int_{t_o}^{s} k(.,\tau,\sigma)\varphi(\tau,\sigma)\ d\tau d\sigma \qquad \varphi \in U^*_{ss} \quad , \tag{4.5}$$

has finite rank. (i.e. it is a finite dimensional operator mapping U^*_{ss} *into* Y).

proof. The proof is immediate because the subspace $\{\mathrm{Im}\ f^*_s\}$ of Y_s is isomorphic to Σ_s via the map H_s. In fact the explicit consideration of $\{\mathrm{Im}\ f^*_s\}$ as the range space of f^*_s leads f^*_s to be surjective. As the diagram (4.4) commutes, H_s is surjective too.

THEOREM. *The linear map* $f^*_s : U^*_{ss} \longrightarrow Y_s$, *where*

$$f^*_s\ (\varphi) = \int_{t_o}^{s} \int_{t_o}^{s} k\ (.,\tau,\sigma)\ \varphi\ (\tau,\sigma)\ d\tau\ d\sigma, \qquad \varphi \in U^*_{ss} \tag{4.6}$$

has finite rank (is finite dimensional) iff there exist two families of continuous maps:

$$\underline{\alpha}\ (.,s) \quad : \ [s,\infty) \longrightarrow R^n \tag{4.7}$$

$$s \in T$$

$$\underline{\beta}\ (s,.,.) \quad : \ [t_o,s] \times [t_o,s] \longrightarrow R^n \tag{4.8}$$

such that the kernel $k(.,.,.)$ *can be factored in the following way:*

$$k(t,\tau,\sigma) = \sum_{i=1}^{n} \alpha_i\ (t,s)\ \beta_i\ (s,\tau,\sigma) \qquad t \in [s,\infty) \quad (\tau,\sigma) \in [t_o s] \times [t_o,s] \tag{4.9}$$

proof. We prove only the necessity since the sufficiency is trivial. The proof of the necessity consists in showing that if

$$\dim \{\mathrm{Im}\ f_s^*\} = n \tag{4.10}$$

then the kernel $k(t,\tau,\sigma)$ can be factored as in (4.9).

Since we allow our present time s to vary, we are actually considering a family of maps like (4.6), depending on the parameter s. As a consequence the factorization to be obtaind will be s-dependent.

Note that no guarantee exists that $\dim\{\mathrm{Im}\ f_s^*\}$ will remain constant as s varies. Actually n should be considered as an integer valued function describing the dimension of the Image space of the map f_s^*. Anyway it is well known that it is always possible to extract from the time set T an open interval J, containing s, where $\dim\{\mathrm{Im}\ f_s^*\}$ is constant.

According to (4.10) let us suppose now that $\{\mathrm{Im}\ f_s^*\}$ is spanned by a set of n linearly independent functions:

$$\alpha_1(.\ s),\dots,\ \alpha_n(.,s)\ ,\ \alpha_i(.,s) : [s,\infty) \longrightarrow R^1 \qquad i=1,\dots n \tag{4.11}$$

so that any $y \in \{\mathrm{Im}\ f_s^*\}$ can uniquely be expressed as:

$$y(.) = \sum_{i=1}^{n} \eta_i\, \alpha_i\, (.,s) \qquad \eta_i \in R^1 \quad i = 1,\dots n \tag{4.12}$$

From Lemma 4 we know that Σ_s is finite dimensional and by isomorphism it has the same dimension as $\{\mathrm{Im}\ f_s^*\}$, $(s \in J)$.

Consider then a basis for Σ_s :

$$\beta_1(s,.,.),\dots,\ \beta_n(s,.,.) \qquad \beta_i(s,.,.) \in U_{ss}^* \quad i = 1,\dots n. \tag{4.13}$$

so that any element belonging to Σ_s has a unique representation

$$\varphi_1(.,.) = \sum_{i=1}^{n} \xi_i \beta_i\ (s,.,.) \qquad \xi_i \in R^1 \qquad i=1,\dots n. \tag{4.14}$$

If the restriction of f_s^* to Σ_s is considered as

$$f_s^* \mid \Sigma_s\ : \Sigma_s \longrightarrow \{\mathrm{Im}\ f_s^*\} \tag{4.15}$$

then there exists an nxn nonsingular matrix $F(s)$, $(s\in J)$ such that:

$$\underline{\eta} = F(s)\, \underline{\xi}\quad ,\quad \underline{\eta} = \mathrm{col}(\eta_1,\dots\eta_n)\quad ,\quad \underline{\xi} = \mathrm{col}\ (\xi_1,\dots,\xi_n). \tag{4.16}$$

The matrix $F(s)$ maps the coordinates $\underline{\xi}$ of $\varphi_1 \in \Sigma_s$ with respect to the basis (4.13) into the corresponding coordinates $\underline{\eta}$ of $f_s^*(\varphi_1)$ as element of $\{\text{Im } f^*_s\}$, taken with respect to the basis (4.11):

$$y(.) = f^*_s(\varphi_1) = \underline{\alpha}^T(.,s)F(s)\,\underline{\xi} \quad ; \quad \varphi_1(.,.) = \underline{\xi}^T \underline{\beta}(s,.,.) \tag{4.17}$$

Note that by (4.1) $f^*_s = (f^*_s|_{\ker f^*_s}, f^*_s|_{\Sigma_s})$, consequently the map f^*_s has been represented by means of the coordinates of the projection of any $\varphi \in U^*_{ss}$ on Σ_s.

It is necessary now to compose the map $f^*_s|_{\Sigma_s}$ with the projection map P_s in order to obtain f^*_s, as in the diagram (4.4).

A useful form for the projection map is obtained noting that the space U^*_{ss} has by its nature an inner product space structure.
In fact defining

$$< \varphi, \psi > = \int_{t_o}^{s}\int_{t_o}^{s} \varphi(\tau,\sigma)\,\psi(\tau,\sigma)\,d\tau\,d\sigma \qquad \varphi,\psi \in U^*_{ss} \tag{4.18}$$

we can be easily convinced that the bilinear form (4.18) is well defined on U^*_{ss} and satisfies trivially the axioms for inner product.

Because of this we can orthonormalize the basis (4.13) and continuing to adopt the same symbol $\underline{\beta}$, write down the projection operator: as:

$$\varphi_1 = P_s(\varphi) = \underline{\beta}^T(s,.,.)\int_{t_o}^{s}\int_{t_o}^{s} \underline{\beta}(s,\tau,\sigma)\varphi(\tau,\sigma)\,d\tau\,d\sigma\ , \quad \varphi \in U^*_{ss} \tag{4.19}$$

A few words to justify (4.19) are in order at this point. Firstly the operator defined by (4.19) is idempotent just by the orthonormality of β_i; secondly note that for any φ belonging to U^*_{ss} one has the unique decomposition:

$$\varphi = \varphi_1 + (\varphi - \varphi_1) \quad , \quad \varphi_1 = P_s(\varphi) \tag{4.20}$$

where φ_1 belongs to Σ_s.

Moreover $(\varphi - \varphi_1)$ is orthogonal to Σ_s.

By doing so the coordinate vector $\underline{\xi}$, of $P_s(\varphi)$ is obtained as:

$$\underline{\xi} = \int_{t_o}^{s}\int_{t_o}^{s} \underline{\beta}(s,\tau,\sigma)\,\varphi(\tau,\sigma)\,d\tau\,d\sigma \qquad \varphi \in U^*_{ss} \tag{4.21}$$

Combining the last espression with (4.17) we get the finite dimensional operator f^*_s in the final form:

$$f^*_s(\varphi) = \underline{\alpha}^T(.,s)\,F(s)\int_{t_o}^{s}\int_{t_o}^{s} \underline{\beta}(s,\tau,\sigma)\,\varphi(\tau,\sigma)\,d\tau\,d\sigma \qquad \varphi \in U^*_{ss} \tag{4.22}$$

The arbitrariness of φ leads immediately to :

$$k(t,\tau,\sigma) = \underline{\alpha}^T(t,s)\ F(s)\ \underline{\beta}(s,\tau,\sigma) \qquad t \in [s,\infty) \qquad (4.23)$$
$$(\tau,\sigma) \in [t_o s] \times [t_o, s]$$

which gives the conclusion once the nonsingularity of F(s) has been re called.

From now on we shall always make the following

Assumption a) : *there exist two families of continuous functions:*

$$\underline{\alpha}(.,s) : [s,\infty) \longrightarrow R^n \qquad (4.24)$$

$$\underline{\beta}(s,.,.) : [t_o s] \times [t_o, s] \longrightarrow R^n \qquad s \in T \qquad (4.25)$$

with linearly independent components, such that :

$$k(t,\tau,\sigma) = \underline{\alpha}^T(t,s)\underline{\beta}(s,\tau,\sigma) \qquad \begin{matrix} s \in [t_o,\infty) \\ t \in [s,\infty) \\ (\tau,\sigma) \in [t_o s] \times [t_o, s] \end{matrix} \qquad (4.26)$$

REMARK. The previous assumption corresponds obviously to hypothyzing constant and finite dimensionality for Σ_s, for any value of the present time parameter s. A practical test for the condition (4.26) to be satisfied seems complicated by the presence of the parameter s. In fact the previous assumption is implied by the following

Assumption b) : *there exist two continuous functions:*

$$\underline{\delta}(.) \ : \ T \longrightarrow R^n \qquad (4.27)$$
$$\underline{\gamma}(.,.) \ : \ T^2 \longrightarrow R^n \qquad (4.28)$$

with linear independent components, such that:

$$k(t,\tau,\sigma) = \underline{\delta}^T(t)\ \gamma(\tau,\sigma) \quad , \quad (t,\tau,\sigma) \in (T \times T^2)^C \qquad (4.29)$$

In order to justify the statement let us define the following truncated functions:

$$\underline{\delta}_s(.) : [s,\infty) \longrightarrow R^n \quad ; \quad \underline{\delta}_s(t) = \begin{cases} \underline{\delta}(t), & t \geqslant s \\ \underline{0}, & t < s \end{cases} \qquad (4.30)$$

$$\underline{\gamma}_s(.,.) : [t_o,s]\times[t_o,s] \longrightarrow R^n \ ; \ \underline{\gamma}_s(\tau,\sigma) = \begin{cases} \underline{\gamma}(\tau,\sigma), & (\tau,\sigma)\in[t_o,s]\times[t_o,s] \\ \underline{0}, & \max(\tau,\sigma) > s \end{cases} \qquad (4.31)$$

By means of (4.30), (4.31) one derives immediately:

$$k(t,\tau,\sigma) = \underline{\delta}_s^T(t)\ \underline{\gamma}_s(\tau,\sigma) \qquad \begin{matrix} s \in T \\ (t,\tau,\sigma) \in (T \times T^2)^C \end{matrix} \qquad (4.32)$$

which holds in the same region as (4.26).

If T(.) is any continuous nonsingular (nxn) matrix, the following relation still holds

$$k(t,\tau,\sigma) = \underline{\delta}_s^T(t)\; T(s)\;.\quad T^{-1}(s)\underline{\gamma}_s(\tau,\sigma) \tag{4.33}$$

Notice that this relation has the same structure as (4.26).

A first consequence of assumption a) (or b)) is a representation for the projection map from U^*_{ss} onto Σ_s. We have in fact:

$$P_s(\varphi) = \underline{\beta}^T(s,.,.) \int_{t_o}^{s}\int_{t_o}^{s} \underline{\beta}(s,\tau,\sigma)\varphi(\tau,\sigma)\, d\tau\, d\sigma \qquad \varphi \in U^*_{ss} \tag{4.34}$$

where we implicitely assumed the orthonormality of $\underline{\beta}(s,.,.)$.

Note that the representation remains valid whatever will be the present time s.

We are now in the condition of combining our preceding results (3.51), (3.52), into an explicit specification of the state equation for the input - output map described by the kernel (4.26).

Assume $t_1 \geqslant s$ and the corresponding canonical factorization of $f^*_{t_1}$ through Σ_{t_1} :

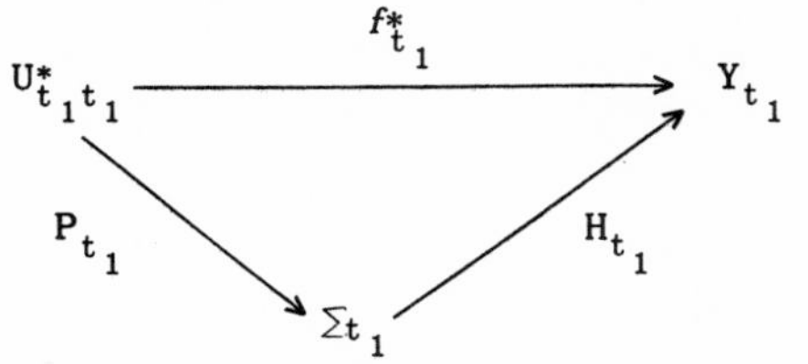

(4.35)

Using the well known decomposition :

$$U^*_{t_1t_1} = U^*_{ss} \oplus U^*_{s,st_1} \oplus U^*_{st_1,s} \oplus U^*_{st_1,st_1} \tag{4.36}$$

we can once more split P_{t_1} by means of the restrictions to the preceding subspaces:

$$\begin{array}{l} U^*_{ss} \xrightarrow{P'_{t1}} \Sigma_{t_1} \\ U^*_{s,st_1} \xrightarrow{P''_{t1}} \Sigma_{t_1} \\ U^*_{st_1,s} \xrightarrow{P'''_{t1}} \Sigma_{t_1} \\ U^*_{st_1,st_1} \xrightarrow{P^{IV}_{t1}} \Sigma_{t_1} \end{array} \tag{4.37}$$

By linearity of P_{t_1} one gets immediately :

$$P_{t_1}(\varphi_{t_1,t_1}) = P'_{t_1}(\varphi_{ss}) + P''_{t_1}(\varphi_{s,st_1}) + P'''_{t_1}(\varphi_{st_1,s}) + P^{IV}_{t_1}(\varphi_{st_1,st_1}) \tag{4.38}$$

where φ_{ss}, φ_{s,st_1}, $\varphi_{st_1,s}$, φ_{st_1,st_1} are the projections of $\varphi_{t_1t_1}$ onto the subspaces U^*_{ss}, U^*_{s,st_1}, $U^*_{st_1,s}$, $U^*_{st_1,st_1}$ respectively.

The first restriction is now factored through Σ_s

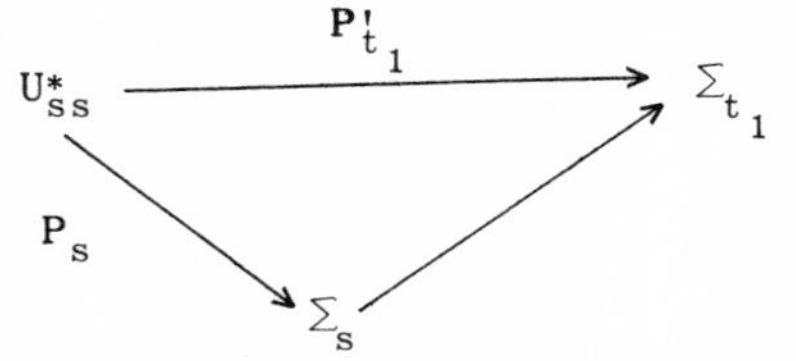

(4.39)

in order to evidentiate an updating structure based on the summarizing action on the past information explicated by Σ_s.

We now examine in more detail the factorization expressed by the diagram (4.39). At this purpose we state the following

LEMMA 5. *Let the families of continuous maps* $\underline{\alpha}(.,s)$ *and* $\underline{\beta}(s,.,.)$ *be as in Assumption* a). *Suppose that* $\underline{\beta}(s,.,.)$ *has orthonormal components, then there exists a real matrix function*

$$\Phi(.,.) : T \times T \longrightarrow (n \times n) \text{ matrices} \tag{4.40}$$

such that

$$\underline{\beta}(t_1,\tau,\sigma) = \Phi(t_1,s)\,\underline{\beta}(s,\tau,\sigma) \qquad (\tau,\sigma) \in [t_o,s] \times [t_o,s] \tag{4.41}$$

The matrix Φ *enjoys the following properties :*

$$\begin{aligned} &a) \quad \Phi(t_2,s) = \Phi(t_2,t_1)\,\Phi(t_1,s) && t_2 \geq t_1 \geq s \\ &b) \quad \Phi(s,s) = I && s \in T \\ &c) \quad \Phi^{-1}(t_1,s) = \Phi(s,t_1) && \end{aligned} \tag{4.42}$$

proof. By assumption a) we can write :

$$k(t,\tau,\sigma) = \underline{\alpha}^T(t,s)\underline{\beta}(s,\tau,\sigma) \quad , \quad t \in [s,\infty) \ , \ (\tau,\sigma) \in [t_o,s] \times [t_o,s] \tag{4.43}$$

$$k(t,\tau,\sigma) = \underline{\alpha}^T(t,t_1)\,\underline{\beta}(t_1,\tau,\sigma) \ , \ t \in [t_1,\infty), \ (\tau,\sigma) \in [t_o,t_1] \times [t_o,t_1] \tag{4.44}$$

this implies:

$$\underline{\alpha}^T(t,s)\underline{\beta}(s,\tau,\sigma) = \underline{\alpha}^T(t,t_1)\underline{\beta}(t_1,\tau,\sigma) \qquad \begin{aligned} &t \in [t_1,\infty) \\ &(\tau,\sigma) \in [t_o,s] \times [t_o,s] \end{aligned} \tag{4.45}$$

for all points (t,τ,σ) belonging to the intersection of the regions where (4.43) and (4.44) are valid.

For a fixed $t \geqslant t_1$, (4.45) can be rewritten as:

$$\underline{\alpha}^T(t,s)\underline{\beta}(s,.,.) = \underline{\alpha}^T(t,t_1)\ \underline{\beta}(t_1,.,.) \tag{4.46}$$

This relationship can be interpreted in the following way: the right hand side must belong to the linear manifold (the n-dimensional space Σ_s) spanned by $\beta_1(s,.,.),\ldots,\beta_n(s,.,.)$. Let us now compute the coordinate vector of $\underline{\alpha}^T(t,t_1)\underline{\beta}(t_1,.,.)$ with respect to this basis. By orthonormality of $\underline{\beta}(s,.,.)$ one has immediately:

$$\int_{t_o}^{s}\int_{t_o}^{s}\underline{\beta}(s,\tau,\sigma)\ \underline{\alpha}^T(t,t_1)\underline{\beta}(t_1,\tau,\sigma)\,d\tau d\sigma = \int_{t_o}^{s}\int_{t_o}^{s}\underline{\beta}(s,\tau,\sigma)\underline{\beta}^T(t_1,\tau,\sigma)\,d\tau d\sigma \cdot$$

$$\cdot\ \underline{\alpha}(t,t_1) \tag{4.47}$$

In view of (4.46) this coordinate vector must equal $\underline{\alpha}(t,s)$. So we can write :

$$\underline{\alpha}(t,s) = \left\{\int_{t_o}^{s}\int_{t_o}^{s}\underline{\beta}(s,\tau,\sigma)\underline{\beta}^T(t_1,\tau,\sigma)\,d\tau\ d\sigma\right\}\underline{\alpha}(t,t_1) \tag{4.48}$$

Define now the n x n matrix :

$$\Phi(s,t_1) = \int_{t_o}^{s}\int_{t_o}^{s}\underline{\beta}(s,\tau,\sigma)\ \underline{\beta}^T(t_1,\tau,\sigma)\ d\tau\ d\sigma \tag{4.49}$$

Reminding that the time instant t was arbitrarily selected (provided that $t \geqslant t_1$), we can rewrite (4.48) as :

$$\underline{\alpha}(.,s) = \Phi(s,t_1)\ \underline{\alpha}(.,t_1) \tag{4.50}$$

The same arguments leading to (4.50) could be repeated for a new value, $t_2 \geqslant t_1$ of the present time. In this case one obtains

$$\underline{\alpha}(.,t_1) = \Phi(t_1,t_2)\ \underline{\alpha}(.,t_2) \tag{4.51}$$

which holds in $[t_2,\infty)$ and, similarly:

$$\underline{\alpha}(.,s) = \Phi(s,t_2)\ \underline{\alpha}(.,t_2) \tag{4.52}$$

These relationships imply the **semigroup property** for Φ :

$$\Phi(s,t_2) = \Phi(s,t_1)\ \Phi(t_1,t_2) \tag{4.53}$$

which follows immediately by linear independence of $\underline{\alpha}(.,t_2)$.

In fact combining (4.50) and (4.51) one gets :

$$\underline{\alpha}(.,s) = \Phi(s,t_1)\ \Phi(t_1,t_2)\ \underline{\alpha}(.,t_2) \tag{4.54}$$

and subtracting (4.54) from 4.52) :

$$[\Phi(s,t_2) - \Phi(s,t_1)\Phi(t_1,t_2)]\ \underline{\alpha}(.,t_2) = \underline{0}\quad . \tag{4.55}$$

Let us now fix an element $(\tau,\sigma)\in[t_o,s]\times[t_o,s]$, then :

$$\underline{\alpha}^T(.,s)\ \mathcal{B}(s,\tau,\sigma) = \underline{\alpha}^T(.,t_1)\ \mathcal{B}(t_1,\tau,\sigma) \tag{4.56}$$

By substituting (4.50) into (4.56) we obtain :

$$\underline{\alpha}^T(.,t_1)\ \Phi^T(s,t_1)\mathcal{B}(s,\tau,\sigma) = \underline{\alpha}^T(.,t_1)\ \mathcal{B}(t_1,\tau,\sigma) \tag{4.57}$$

Invoking once more the linear independency of the components of $\underline{\alpha}(.,t_1)$, we get:

$$\mathcal{B}(t_1,\tau,\sigma) = \Phi^T(s,t_1)\ \mathcal{B}(s,\tau,\sigma) \tag{4.58}$$

$$(\tau,\sigma)\ \varepsilon\ [t_o,s]\ x\ [t_o,s]$$

An explicit expression for the transpose of the matrix $\Phi(s,t_1)$ may be derived taking the transpose of both members of (4.49) :

$$\Phi^T(s,t_1) = \int_{t_o}^{s}\int_{t_o}^{s}\mathcal{B}(t_1,\tau,\sigma)\ \mathcal{B}^T(s,\tau,\sigma)\ d\tau\ d\sigma = \Phi(t_1,s) \tag{4.59}$$

In this way the previous relationship allows to extend the domain of definition of the matrix $\Phi(.,.)$ to the region where the first argument takes values bigger than the second one.

Using (4.59) we can write out the main result:

$$\mathcal{B}(t_1,\tau,\sigma) = \Phi(t_1,s)\ \mathcal{B}(s,\tau,\sigma),\quad (\tau,\sigma)\ \varepsilon\ [t_o,s]\ x\ [t_o,s] \tag{4.60}$$

which represents the updating formula for the basis $\mathcal{B}(s,.,.)$.

The combination of (4.53) and (4.59) proves directly property a). The property b) follows from (4.59) simply putting $t_1 = s$.

Only property c) has still to be proved. To this purpose put t_2=s in (4.53) :

$$I = \Phi(s,s) = \Phi(s,t_1)\ \Phi(t_1,s) \tag{4.61}$$

In view of (4.59) one has:

$$I = \Phi(s,t_1)\ \Phi^T(s,t_1) \tag{4.62}$$

from which it is apparent that $\Phi(s,t_1)$ is nonsingular because $1 = (\det\Phi(s,t_1))^2$. Q. E. D.

As an immediate consequence of Lemma 5 one can write out explicitely the factorization of $\mathbf{P}'_{t_1}$ represented by the diagram (4.39). To this purpose we note that:[1]

$$P'_{t_1}(\varphi_{ss}) = \mathcal{B}^T(t_1,.,.)\int_{t_o}^{s}\int_{t_o}^{s}\mathcal{B}(t_1,\tau,\sigma)\ \varphi_{ss}(\tau,\sigma)\ d\tau\ d\sigma\ ,\qquad \varphi_{ss}\varepsilon U^*_{ss} \tag{4.63}$$

Invoking (4.41), the previous relationship becomes:

$$P'_{t_1}(\varphi_{ss}) = \mathcal{B}^T(t_1,.,.)\Phi(t_1,s)\int_{t_o}^{s}\int_{t_o}^{s}\mathcal{B}(s,\tau,\sigma)\ \varphi_{ss}(\tau,\sigma)\ d\tau\ d\sigma \tag{4.64}$$

Observe that the integral in the right hand side of (4.64) represents the coordinate vector of $P_s(\varphi_{ss})$.

Since every n-dimensional vector space is isomorphic to R^n, we can refer to (4.64) as the composition of maps represented in the commutative diagram

$$\begin{array}{ccccc} U^*_{ss} & & \xrightarrow{\quad P'_{t_1} \quad} & & \Sigma_{t_1} \\ \downarrow P_s & & & & \uparrow \text{isomorphism} \\ \Sigma_s & \xrightarrow[\text{isomorphism}]{} & R^n & \xrightarrow[\Phi(t_1, s)]{} & R^n \end{array} \qquad (4.65)$$

In other words the factorization (4.39) turns out to be canonical.

We shall now attempt to draw some consequences of the previous considerations by means of the following

THEOREM. *Let* (4.26) *hold. Suppose that the limit :*

$$A(s) = \lim_{h \to 0} \frac{1}{h} [\Phi(s+h, s) - I] \qquad h \in R_+ \qquad (4.66)$$

exists for almost all $s \in T$. *Let* $A(.)$ *be a measurable* $n \times n$ *matrix function. Then the input-output pairs generated by the bilinear map* (2.10) *can be constructed linearly transforming the solutions of the differential equation :*

$$\dot{\underline{x}}(t) = A(t)\underline{x}(t) + u(t) \int_{t_o}^{t} \underline{\beta}(t, t, \sigma) v(\sigma) d\sigma + v(t) \int_{t_o}^{t} \underline{\beta}(t, \tau, t) u(\tau) d\tau \qquad (4.67)$$

$$t \geqslant t_o$$

starting from $\underline{x}(t_o) = \underline{0}$. *The linear transformation is expressed by the readout map :*

$$y(t) = \underline{\alpha}^T(t, t)\ \underline{x}(t) \qquad t \geqslant t_o \quad (4.68)$$

proof. We prove the Theorem in four steps.

Step 1. Compose the projection maps P''_{t_1}, P'''_{t_1}, $P^{IV}_{t_1}$ with the tensor product $\otimes$. Recalling the general expression for the projection onto Σ_{t_1}, we obtain :

$$P''_{t_1}(u \cdot w) = \underline{\beta}^T(t_1, ., .) \int_{t_o}^{s} \int_{s}^{t_1} \underline{\beta}(t_1, \tau, \sigma)\ u(\tau)\ w(\sigma)\ d\tau\ d\sigma \qquad (4.69)$$

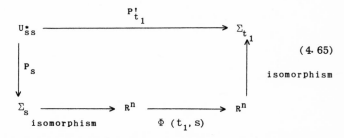

$$P'''_{t_1}(z\cdot v) = \mathcal{B}^T(t_1,\ldots) \int_s^{t_1}\int_{t_o}^{s} \mathcal{B}(t_1,\tau,\sigma)\ z(\tau)\ v(\sigma)\ d\tau\ d\sigma \tag{4.70}$$

$$P^{IV}_{t_1}(z\cdot w) = \mathcal{B}^T(t_1,\ldots)\int_s^{t_1}\int_s^{t_1} \mathcal{B}(t_1,\tau,\sigma)\ z(\tau)\ w(\sigma)\ d\tau\ d\sigma \tag{4.71}$$

$$u, v \in U_s\ , \qquad z, w \in U_{st_1}$$

Step 2. On the basis of (4.64), (4.69), (4.70) and (4.71) rewrite (4.38) for an element of the manifold M_t (see definition (3.35)). Instead of referring directly to elements of Σ_{t_1}, we can simply write out the corresponding relationship for their coordinates, i.e.:

$$\underline{x}(t_1) = \Phi(t_1,s)\underline{x}(s) + \int_{t_o}^{s}\int_s^{t_1} \mathcal{B}(t_1,\tau,\sigma)u(\tau)w(\sigma)d\tau\ d\sigma +$$

$$+ \int_s^{t_1}\int_{t_o}^{s} \mathcal{B}(t_1,\tau,\sigma)z(\tau)v(\sigma)d\tau\ d\sigma +$$

$$+ \int_s^{t_1}\int_s^{t_1} \mathcal{B}(t_1,\tau,\sigma)z(\tau)w(\sigma)\ d\tau\ d\sigma\ . \tag{4.72}$$

where :

$$\underline{x}(t_1) = \int_{t_o}^{t_1}\int_{t_o}^{t_1} \mathcal{B}(t_1,\tau,\sigma)\ u^*(\tau)\ v^*(\sigma)\ d\tau\ d\sigma \tag{4.73}$$

$$\underline{x}(s) = \int_{t_o}^{s}\int_{t_o}^{s} \mathcal{B}(s,\tau,\sigma)\ u(\tau)\ v(\sigma)\ d\tau\ d\sigma \tag{4.74}$$

$$u^* = u + z \in U_{t_1}$$

$$v^* = v + w \in U_{t_1}$$

Step 3. If the limit (4.66) exists, then the following derivative exists too:

$$\frac{d}{dt_1}\underline{x}(t_1) = \frac{d}{dt_1}\Big\{\Phi(t_1,s)\Big[\underline{x}(s) + \int_{t_o}^{s}\int_s^{t_1}\mathcal{B}(s,\tau,\sigma)\ u(\tau)\ w(\sigma)\ d\tau\ d\sigma +$$

$$+ \int_s^{t_1}\int_{t_o}^{s}\mathcal{B}(s,\tau,\sigma)\ z(\tau)v(\sigma)d\tau d\sigma + \int_s^{t_1}\int_s^{t_1}\mathcal{B}(s,\tau,\sigma)z(\tau)w(\sigma)d\tau\ d\sigma\Big]\Big\} \tag{4.75}$$

After some simple manipulations we recognize that this derivative gives precisely (4.67).

Note that $\beta(t,t,\sigma)$ and $\beta(t,\tau,t)$ are well defined as

$$\lim_{\tau\uparrow t} \beta(t,\tau,\sigma) \qquad \sigma \leqslant t \tag{4.76}$$

and

$$\lim_{\sigma\uparrow t} \beta(t,\tau,\sigma) \qquad \tau \leqslant t \tag{4.77}$$

respectively.

Step 4. If (4.26) is valid, then $\underline{\alpha}(.,s)$ can always be assumed as a basis for the Image space of the map f_s^*. Then, just by definition of the tensor product (see (3.36)), it turns out that $\underline{\alpha}(.,s)$ is a basis for the smallest linear vector space containing $\{\mathrm{Im}\ f_s\}$. Because of this we can write

$$y(.) = \underline{\alpha}^T(.,s)\ \underline{x}(s) \tag{4.78}$$

where $y(.) \in Y_s$ is any (free) response in the range of the map f_s.

The value of the output at time s is therefore

$$y(s) = \underline{\alpha}^T(s,s)\ \underline{x}(s) \qquad s \geqslant t_o. \tag{4.79}$$

At this point we can conclude our proof simply referring to the definition (3.33)

REMARK. Note that the map defined by (4.72), which can be regarded as the solution of the differential equation (4.67) starting from initial state $\underline{x}(s)$, depends on the entire past history of the input. This is apparent from its dependency on u and v.

A further investigation is in order at this point to know how finite dimensional representations can be extracted from:

$$\int_{t_o}^{t} \beta(t,t,\sigma)\ v(\sigma)\ d\sigma \quad , \quad \int_{t_o}^{t} \beta(t,\tau,t)\ u(\tau)\ d\tau\ , \tag{4.80}$$

but it is apparent that we can refer to (4.80) as linear input-output maps. At this point the problem may be simply reduced to the realization problem for linear input-output maps.

Define the family of linear maps :

$$f_1 : U_s \longrightarrow Y_s^n \quad , \quad f_1(u) = \int_{t_o}^{s} \beta(,\tau,t)\ u(\tau)\ d\tau \qquad t \geqslant s \tag{4.81}$$

$$f_2 : U_s \longrightarrow Y_s^n \quad , \quad f_2(v) = \int_{t_o}^{s} \beta(,t,\sigma)\ v(\sigma)\ d\sigma \qquad t \geqslant s \tag{4.82}$$

and recognize that they are the "free" counter parts of ii) and iii) in section 3.

A finite dimensional realization for f_1 and f_2 is obtainable proceeding as in the previous derivation (or following more classical methods, see [5]). More specifically we have the following

PROPOSITION. *Let* f_i, $i = 1, 2$, *be defined by* (4.81) *and* (4.82). *The input-output maps* $f_i, i=1,2$ *admit a finite dimensional realization iff two continuous matrix functions*

$$Q_1(.,s) : [s,\infty) \longrightarrow (n \times n_1) \text{ matrices} \quad (4.83)$$

$$s \in T$$

$$Q_2(.,s) : [s,\infty) \longrightarrow (n \times n_2) \text{ matrices} \quad (4.84)$$

and two continuous vector valued functions :

$$\underline{\omega}_1(s,.) : [t_o, s] \longrightarrow R^{n_1} \quad (4.85)$$

$$s \in T$$

$$\underline{\omega}_2(s,.) : [t_o, s] \longrightarrow R^{n_2} \quad (4.86)$$

exist, such that :

$$\beta(t,\tau, t) = Q_1(t,s)\ \underline{\omega}_1(s,\tau) \qquad t \in [s,\infty),\ \tau \in [t_o, s] \quad (4.87)$$

$$\beta(t,t,\sigma) = Q_2(t,s)\ \underline{\omega}_2(s,\sigma) \qquad t \in [s, \infty),\ \sigma \in [t_o, s] \quad (4.88)$$

for any $s \in T$.

Assume that (4.87) and (4.88) hold. This allows us to refer directly to the following state space descriptions

$$\begin{aligned} \dot{\underline{x}}_1(t) &= A_1(t)\ \underline{x}_1(t) + \underline{b}_1(t)\ u(t) \\ \underline{y}_1(t) &= Q_1(t,t)\ \underline{x}_1(t) \end{aligned} \qquad t \geqslant t_o \quad (4.89)$$

$$\begin{aligned} \dot{\underline{x}}_2(t) &= A_2(t)\ \underline{x}_2(t) + \underline{b}_2(t)\ v(t) \\ \underline{y}_2(t) &= Q_2(t,t)\ \underline{x}_2(t) \end{aligned} \qquad t \geqslant t_o \quad (4.90)$$

which provide the finite dimensional realizations of (4.81) and (4.82) we were asking for.

We quote here the expressions for $A_1(t)$, $A_2(t)$, $\underline{b}_1(t)$ and $\underline{b}_2(t)$, which are defined in a well known way :

$$\underline{\omega}_i(t,.) = \Phi_i(t,s)\ \underline{\omega}_i(s,.) \qquad t \geqslant s \ , \quad i = 1,2 \quad (4.91)$$

$$A_i(t) = \lim_{h \downarrow 0} \frac{1}{h} [\Phi_i(t+h,t) - I] \qquad t \geqslant t_o \quad i = 1,2 \quad (4.92)$$

$$\underline{b}_1(t) = \lim_{\tau \uparrow t} \underline{\omega}_1(t,\tau) \quad , \quad \underline{b}_2(t) = \lim_{\sigma \uparrow t} \underline{\omega}_2(t,\sigma) \quad (4.93)$$

At this point a significant conclusion can be drawn. We summarize it in the following

PROPOSITION. *Assume that* (4.26) *holds. Suppose in addition that* $\beta(s,.,.)$ *satisfies* (4.87) *and* (4.88). *Then the bilinear input-output map* (2.10) *is realized by the dynamical system described by the composite state equations :*

$$\dot{\underline{x}}(t) = A(t)\,\underline{x}(t) + u(t)\,\underline{y}_2(t) + v(t)\,\underline{y}_1(t) \qquad \underline{x}(t_o) = \underline{0}$$

$$\dot{\underline{x}}_1(t) = A_1(t)\,\underline{x}_1(t) + \underline{b}_1(t)\,u(t) \qquad \underline{x}_1(t_o) = \underline{0}$$

$$\underline{y}_1(t) = Q_1(t,t)\,\underline{x}_1(t)$$

$$\dot{\underline{x}}_2(t) = A_2(t)\,\underline{x}_2(t) + \underline{b}_2(t)\,v(t)$$

$$\underline{x}_2(t_o) = \underline{0}$$

$$\underline{y}_2(t) = Q_2(t,t)\,\underline{x}_2\,(t)$$

(4.94)

and the associate readout map :

$$y(t) = \underline{\alpha}^T(t,t)\;\underline{x}(t) \qquad (4.95)$$

References

(1) A. V. Balakrishnan, "On the State Space Theory of Nonlinear Systems", in Functional Analysis and Optimization, R. Caianello ed. Academic Press, N.Y. 1966.

(2) M. A. Arbib, "A Characterization of Multilinear Systems", IEEE, T-AC (short paper), Vol AC 14, pp. 699-702, Dec. 1969.

(3) M. Schetzen, "Synthesis of a Class of Nonlinear Systems", Q.P.R., No. 68, Res. Lab. Elec. M.I.T., Jan. 1964.

(4) A.M. Bush, "Kernels Realizable Exacly with a Finite Number of Linear Systems and Multipliers", Q.P.R., No 76, Res. Lab. Elec. M.I.T., 1965.

(5) R.E. Kalman, P.L. Falb, M.A. Arbib, "Topics in Mathematical Systems Theory", McGraw-Hill, 1969.

(6) J.C. Willems, S.K. Mitter, "Controllability, Observability, Pole Allocation and State Reconstruction", IEEE, T-AC, Vol. AC 16, pp. 582-595, Dec. 1971.

(7) A. Nerode, "Linear Automaton Transformations", Proc. Am. Math. Soc., Vol. 9, pp. 541-544, 1958.

(8) S. Lang, "Algebra", Addison Wesley Pub. Co., 1965.

MATHEMATICAL MODELS AND IDENTIFICATION OF BILINEAR SYSTEMS

C. Bruni*- G. Di Pillo**-G. Koch*

Abstract

In this paper the authors present for the class of bilinear systems a dynamic input-state-output model which is valid in a stochastic sense. Then, as for as identification is concerned, the problem is dealt with under deterministic hypotheses applying Newton's method and under stochastic hypotheses using the approach of maximum likelihood estimation.

Introduction

The systems we deal with are called bilinear since they present in their differential input-state model a non-linearity introduced by products between state variables and input variables. They play a significant role from an applicative point of view, since they can describe several important processes in physics, biology and economics [1,2]. These systems have been studied especially in connection with optimization, controllability [1,3] and modeling [4] problems. In this paper, as far as modeling is concerned, following some results of Mc Shane [13,14], a dynamic input-state-output model will be presented which is valid in a stochastic sense. As far as identification is concerned, the problem is dealt with under deterministic hypotheses, applying Newton's method. Then we allow for statistical hypotheses using the stochastic model previously defined and we follow the approach of Maximum Likelihood Estimation.

In both cases, the aim is to determine a vector $b \in R^r$ of unknown parameters which appear in the input-state differential equation, and the state $x(t) \in R^n$, $\forall t$, given input

* C.Bruni and G.Koch are with the Istituto di Automatica, Università di Roma, Roma (Italy)

** G.Di Pillo is with the CSSCCA, C.N.R. (National Council of Researches), Roma (Italy)

This work has been done with support of C.N.R.

and output measurements in the time interval $[0,T]$.

In order to formulate the problem in a unified way, let us define the *augmented state vector:*

$$z(t) \triangleq \begin{bmatrix} x(t) \\ b \end{bmatrix} \in R^m, \forall t \in [0,T], \quad m = r + n \tag{1.1}$$

Now, since from (1.1) we get

$$\dot{z}(t) = \begin{bmatrix} \dot{x}(t) \\ 0 \end{bmatrix} \tag{1.2}$$

from the bilinear model a new differential model can be deduced for the augmented state (1.1).

Thus, the identification problem is reduced to that of determining the state of the newly introduced model.

Identification of bilinear systems under deterministic hypotheses

Problem formulation

A continuous bilinear system may be described by the following differential model [2,3,4]:

$$\dot{x}(t) = Ax(t) + Bx(t)u(t) + Cu(t) \tag{2.1}$$

$$y(t) = Dx(t) \tag{2.2}$$

where $x(t)$, $u(t)$, $y(t)$ are respectively the state, input, output vectors in R^n, R^p, R^q $\forall t \in [0,T]$. A, C, D are constant matrices of suitable dimensions. B is a bilinear operator which maps the space $R^n \times R^p$ into the space R^n; it can be represented as a tri-dimensional $(n \times n \times p)$ constant matrix with elements b_{ijk} in such a way that

$$Bx(t)u(t) \triangleq \begin{bmatrix} \sum_{1}^{p}{}_k \sum_{1}^{n}{}_j \; b_{1jk} x_j(t) u_k(t) \\ \cdots\cdots\cdots\cdots\cdots\cdots \\ \sum_{1}^{p}{}_k \sum_{1}^{n}{}_j \; b_{njk} x_j(t) u_k(t) \end{bmatrix} \tag{2.3}$$

Let us further assume u is measurable and bounded in norm by an integrable function on $[0,T]$. Under these assumptions, for any $x(0) = \hat{x}$, a unique solution of (2.1) exists [5].

Introducing the augmented state z defined by (1.1), from the bilinear model (2.1), (2.2) we deduce the model

$$\dot{z}(t) = Q(t)z(t)z(t) + L(t)z(t) + c(t) \tag{2.4}$$

$$y(t) = Gz(t) \tag{2.5}$$

where $Q(t)$, $\forall t$, is a $m\times m\times m$ matrix which maps the space $R^m \times R^m$ into the space R^m. $L(t)$ and $c(t)$ are respectively a $m\times m$ matrix and a $m\times 1$ vector.

Since $Q(t)$, $L(t)$, $c(t)$ are linear in $u(t)$, they satisfy the same measurability and boundedness assumptions; finally, G is a constant $q\times m$ matrix.

By assumption, we know $y(t)$ without noise in the discrete instants $t_i \in [0,T]$, $i = 1,2, \dots ,N$; we also know u and consequently Q, L, c, G. The identification problem consists in finding the solution of (2.4) with the conditions:

$$y(t_i) = Gz(t_i) \qquad i = 1,2, \dots ,N \tag{2.6}$$

Since we have a deterministic problem, the necessary number of conditions equals the order m of the problem: hence in the following we shall assume $N = m/q$ (*). It will be useful to write conditions (2.6) in a more compact form:

$$\tilde{y} = \sum_{1}^{N}{}_i \; G_i z(t_i) \tag{2.7}$$

where

$$\tilde{y} \triangleq \begin{bmatrix} y(t_i) \\ \cdot \\ \cdot \\ y(t_N) \end{bmatrix} \qquad G_i \triangleq \begin{bmatrix} 0 \\ \cdot \\ \cdot \\ G \\ \cdot \\ \cdot \\ 0 \end{bmatrix} \tag{2.8}$$

(*) When N is not a multiple of q, there are no conceptual differences, but only a few additional formal remarks are required (we can make N equal to the smallest integer greater than m/q).

G_i is a m×m matrix, built with N q×m blocks, all of which are zero except the i-th block equal to G.

Now the deterministic parametric identification problem of bilinear systems has been formulated as the non linear (quadratic) multipoint boundary value problem (MPBVP) (2.4), (2.7).

For the reasons which will be clarified below, it is convenient to transform this problem into an equivalent integral problem.

For that purpose the following theorem has been proved [6]:

Theorem 1

Let V(t) and M_i, i = 1,2, ... ,N be m×m matrices, with V(t) measurable and bounded in norm by an integrable function on $[0,T]$.

Let us denote by $\Phi(t,\tau)$ the solution of the equation:

$$\frac{\partial\Phi(t,\tau)}{\partial t} = V(t)\Phi(t,\tau), \quad \Phi(\tau,\tau) = I, \quad t,\tau\in[0,T] \tag{2.9}$$

Let us define the m×m matrix:

$$P \triangleq \sum_{1}^{N} M_i \Phi(t_i,0) \tag{2.10}$$

Under the assumption that P is nonsingular, the MPBVP defined by (2.4), (2.7) where Q, L, c are measurable and bounded in norm by an integrable function on $[0,T]$, is equivalent to the integral equation:

$$z(t) = R^{(V,M_i)}(t)\left[\tilde{y} - \sum_{1}^{N}(G_i - M_i)z(t_i)\right] +$$

$$+ \int_0^T S^{(V,M_i)}(t,\tau)\{[Q(\tau)z(\tau) + L(\tau) - V(\tau)]z(\tau) +$$

$$+ c(\tau)\}d\tau \triangleq \Gamma(z)(t) \tag{2.11}$$

$R^{(V,M_i)}(t)$, $S^{(V,M_i)}(t,\tau)$ can be considered like Green's matrices for the MPBVP and are defined by

$$R^{(V,M_i)}(t) \triangleq \Phi(t,0)P^{-1} \tag{2.12}$$

$$S^{(V,M_i)}(t,\tau) \triangleq \Phi(t,0)P^{-1} \sum_{1}^{N} {}_i G_i \Phi(t_i,\tau)\left[\delta_{-1}(t-\tau)-\delta_{-1}(t_i-\tau)\right] \tag{2.13}$$

where δ_{-1} is the unit step function. * * *

Thus, Theorem 1 states that the differential MPBVP (2.4), (2.7) is equivalent to the nonlinear integral problem

$$\Psi(z) \triangleq z - \Gamma(z) = 0 \tag{2.14}$$

in the sense that each solution of the first one is a solution of the second and conversely each absolutely continuous solution of the second is a solution of the first.

As far as the choice of the matrix set $\{V(t), M_i\}$ is concerned, so that P is nonsingular, we have the following theorem [6]:

Theorem 2

Let us consider a N-tuple of m×m matrices M_i, with the same structure of G_i in (2.8). A necessary and sufficient condition for the existence of an m×m matrix V(t) such that P is nonsingular, is:

$$\text{rank}\left[M_1 \; M_2 \; \dots \; M_N\right] = m \qquad * * *$$

In the sufficiency part of the proof, a constructive procedure for a possible V(t) is given.

Solution of the problem by means of Newton's method

The operator Ψ defined by (2.14) maps the space Z of absolutely continuous functions in $[0,T]$ into Z itself. The nonlinear equation (2.14) can be numerically solved via iterative algorithms.

In this Section we consider Newton's method [7,8,9], also known as quasilinearization algorithm [10,11]. It consists in generating the sequence

$$z_n = z_{n-1} - \left[\Psi'(z_{n-1})\right]^{-1}\Psi(z_{n-1}) \quad n=1,2, \dots \tag{2.16}$$

starting from a given $z_o \in Z$.

Once Z is equipped with a norm which makes it a Banach space, we can study the convergence of (2.16) by means of the well-known theorem of Kantorovich [7,8,9] which also yields informations about existence and uniqueness of solutions of (2.14) in a neighborhood of z_o. Exploiting the particular form of Ψ in our problem, from the above theorem we get ad hoc conditions, which can be more easily checked.

Actually, one of the assumptions of Kantorovich's theorem particularly difficult to verify is

$$\| \Psi''_{(z)} \| \leq k \tag{2.17}$$

in a closed sphere in Z centered at z_o with radius r sufficiently large. Now for our problem, being

$$\Psi''_{(z)}(u,v)(t) = - \int_o^T S^{(V,M_i)}(t,\tau)Q(\tau)u(\tau)v(\tau)d\tau \tag{2.18}$$

$\Psi''_{(z)}$ is indipendent of z and bounded, so that (2.17) is verified for some k in all Z.

As a result, from the theorem of Kantorovich we get [6]:

Theorem 3

Let us consider the integral equation (2.14) equivalent to the MPBVP (2.4), (2.7). Let:

$$k \geq \| \Psi''_{(z)} \| , \quad z \in Z.$$

If: a) $z_o \in Z$ exists so that:

$$\| \Gamma'_{(z_o)} \| = \| R^{(V,Mi)}(t) \sum_1^N {}_i (M_i - C_i) +$$

$$+ \int_o^T S^{(V,M_i)}(t,\tau) [2Q(\tau)z_o(\tau)+L(\tau)-V(\tau)] d\tau \| < 1^{(*)} \tag{2.19}$$

(*) In (2.19) we implicitly assumed the quadratic operator Q defined by the matrix Q(t) to be symmetric (Qz_1z_2 =

b) defined:

$$\beta \geq \frac{1}{1 - \| \Gamma'_{(z_o)} \|} \tag{2.20}$$

$$\eta \geq \| z_1 - z_o \| = \| [\Psi'_{(z_o)}]^{-1} \Psi(z_o) \| \tag{2.21}$$

one has:

$$h \triangleq \eta\beta k \leq \frac{1}{2} \tag{2.22}$$

then Newton's method applied to the given problem, starting from z_o, converges to the unique solution in the sphere centered at z_o and with radius

$$r_1 \triangleq \frac{1 + \sqrt{1-2h}}{h} \eta \tag{2.23}$$

The sphere should be open if $h < \frac{1}{2}$ and closed if $h = \frac{1}{2}$. * * *

Introducing further assumptions on the MPBVP to be solved, the study of the convergence of Newton's method can be simplified.

In this connection we have the following theorem [6]:

$= Qz_2z_1$) so that:

$$\left.\frac{d(Qzz)}{dz}\right|_{z_o} = 2Qz_o$$

This indeed imposes no restriction since if that were not the case we would have:

$$\left.\frac{d(Qzz)}{dz}\right|_{z_o} = Qz_o + Q^*z_o \triangleq 2\bar{Q}z_o$$

where * indicates transposes and the "average operator" $\bar{Q}$ is symmetric.

Theorem 4

Let us consider the integral equation (2.14) equivalent to the MPBVP (2.4), (2.7). If:

a)

$$\text{rank}\left[G_1 \; G_2 \; \ldots . \; G_N\right] = m \tag{2.24}$$

b) $z_o \in Z$ exists so that with the choice:

$$V(t) = 2Q(t)z_o(t) + L(t) \tag{2.25}$$

$$M_i = G_i \qquad i = 1,2, \; \ldots \; ,N \tag{2.26}$$

the matrix P defined by (2.10) is nonsingular

c)

$$\| \Psi(z_o) \| \leq \frac{1}{2k} \tag{2.27}$$

where Ψ and k are defined by (2.25), (2.26), (2.17), then Newton's method starting from z_o converges to the unique solution in the sphere centered at z_o and with radius:

$$r_1' \triangleq \frac{1 + \sqrt{1-2k\| \Psi(z_o) \|}}{k} \tag{2.28}$$

The sphere is open or closed according to whether in (2.27) we have the inequality or equality sign. * * *

Theorem 4 may be immediately proved from Theorem 3 once we note that because of (2.25), (2.26), from (2.19) we get $\| \Gamma'_{(z_o)} \| = 0$.

As far as computing the constant k (which appears in both Theorem 3 and 4) is concerned, if the norm in the space Z is defined by:

$$\| z \| \triangleq \max_{t \in [0,T]} \; \max_{1 \leq j \leq m} |z_k(t)| \tag{2.29}$$

k can be obtained by the following inequalities:

$$\| \Psi''_{(z)} \| \triangleq \sup_{\substack{\|u\| \leq 1 \\ \|v\| \leq 1}} \left\| \int_o^T S^{(V,M_i)}(t,\tau)Q(\tau)u(\tau)v(\tau)d\tau \right\| =$$

$$= \sup_{\substack{\|u\| \leq 1 \\ \|v\| \leq 1}} \max_{t \in [0,T]} \max_{1 \leq j \leq m} \left| \sum_{1}^{m}{}_h \int_o^T s_{jh}^{(V,M_i)}(t,\tau) \sum_{1}^{m}{}_{rs} q_{hrs}(\tau)u_s(\tau)v_r(\tau)d\tau \right| \leq$$

$$\leq \max_{t \in [0,T]} \max_{1 \leq j \leq m} \sum_{1}^{m}{}_{rs} \int_o^T \left| \sum_{1}^{m}{}_h s_{jh}^{(V,M_i)}(t,\tau)q_{hrs}(\tau) \right| d\tau \triangleq k$$

where $s_{jh}^{(V,M_i)}(t,\tau)$ and $q_{hrs}(t)$ are the elements respectively of matrices $S^{(V,M_i)}(t,\tau)$ and $Q(t)$.

As a final remark, note that the sequence$\{z_k\}$ generated by (2.16) from a given z_o is indipendent of the matrix set $\{V(t), M_i\}$ which defines $^o\Psi$, as is easily verified [6]. In particular, it coincides with the sequence starting from z_o and generated according to:

$$\dot{z}_{n+1}(t) = [2Q(t)z_n(t) + L(t)]z_{n+1}(t) - Q(t)z_n(t)z_n(t) + + c(t) \qquad (2.31)$$

$$\sum_{1}^{N}{}_i G_i z_{n+1}(t_i) = \tilde{y} \qquad (2.32)$$

which are obtained by direct application of Newton's method to the MPBVP (2.4), (2.7).

However, the importance of our procedure, which is based on the integral representation of the MPBVP,stems from the possibility of studying the convergence of the algorithm.

Indeed Theorem 3 and 4 yield sufficient convergence

conditions which are more easily checked than those we may directly obtain for the sequence (2.31), (2.32).

Actually, it can be shown [6] that the linear MPBVP (2.31), (2.32) admits the solution:

$$z_{n+1}(t) = \Gamma^{(z_n)}(z_n)(t) \tag{2.33}$$

where $\Gamma^{(z_n)}$ is defined according to (2.11) and the choice

$$V(t) = 2Q(t)z_n(t) + L(t) \tag{2.34}$$

$$M_i = G_i \tag{2.35}$$

Thus, the study of the convergence of the sequence generated by (2.34) is not straightforward due to the dependence of $\Gamma^{(z_n)}$ on z_n itself.

Identification of bilinear systems under stochastic hypotheses

Modeling stochastic bilinear systems

The differential model (2.1) of the previous Section is not adequate to represent a bilinear system driven by a random input, such as white noise. In fact, in such a case, the state x could turn out to be not differentiable with probability 1 [12].

On the other hand, a random input should be accomodated for wherever input behaviour is not a priori known due to lack of information, or measurement errors.

For that reason, model (2.1) has to be generalized in a stochastic model. Following Mc Shane [13,14] we can show that the stochastic integral equation equivalent to (2.1)is of the form

$$x(t) = \hat{x} + \int_0^t Ax(\tau)d\tau + \int_0^t [Bx(\tau) + C]dr(\tau) +$$

$$+ \frac{1}{2}\int_0^t B\{[Bx(\tau) + C]dr(\tau)\}dr(\tau) \quad , \; t \in [0,T] \tag{3.1}$$

where the random variable $\hat{x} \in R^n$ takes initial conditions into account and the p-component stochastic process r represents the input process. In the RHS of (3.1) the first integral is a standard Lebesgue integral, while the second and the third are respectively defined as simple and double stochastic (Mc Shane) integrals [13,15]. It is noteworthy that (3.1) and (2.1) are equivalent in the sense that if r is Lipschitzian with probability 1 so that it can be written as:

$$r(t) = \int_0^t u(\tau)d\tau + r_o \qquad t \in [0,T] \tag{3.2}$$

with r_o constant and u integrable in $[0,T]$, then in the RHS of (3.1) the second integral reduces to a Lebesgue integral and the third vanishes, so that (3.1) reduces to (2.1).

Moreover, if there is a closed form of the solution x as a function of the input, this form is the same for both equations.

The first problem which arises is connection with (3.1) is the existence and uniqueness of solutions.We can prove the following theorem [16]:

Theorem 5

a) Let (Ω, A, P) be a probability space

b) Let $\{F_t\}$, $t \in [0,T]$, be a family of σ-Algebras on Ω such that $F_t \subset F_\tau$, $t < \tau$, and $F_t \subset A$, $t \in [0,T]$

c) Let $\hat{x}$ be a random variable in R^n, F_o-measurable, whith the first two moments finite

d) Let r be a stochastic process defined in $[0,T]$ on the probability space, so that:

d1) $r(t)$ is F_t-measurable, $\forall t \in [0,T]$

d2) three finite constants exist, k_1, k_2, k_4, so that:

$$E\left[\| r(t) - r(t) \|^i \middle| F_\tau \right] \le k_i |t-\tau| \qquad i = 1,2,4 \quad 0 \le \tau < t \le T \tag{3.3}$$

Then there exists a solution x of (3.1), unique with probability 1, continuous and F_t-measurable $\forall t \in [0,T]$, and we have:

$$E\left[\| x(t) - x(\tau) \|^{i}\right] \leq k_i' |t-\tau| \qquad \begin{array}{l} i = 1, 2 \\ 0 \leq \tau < t \leq T \end{array} \tag{3.4}$$

with k_1' , k_2' finite constants. * * *

The second problem is to find a numerical procedure to compute an approximate solution of (3.1). We introduce the *Cauchy polygonal approximation:* let us consider the partition Π of the interval $[0,T]$:

$$\Pi = \{t_1, t_2, \ldots , t_{k+1}\} \; ; \quad 0 = t_1 < t_2 < \ldots < t_{k+1} = T \tag{3.5}$$

and define the quantity:

$$m(\Pi) = \max_{i=1,2 \ldots k} |t_{i+1} - t_i| \tag{3.6}$$

The Cauchy polygonal approximation x_π is defined by:

$$x_\pi(0) = \hat{x} \tag{3.7}$$

$$x_\pi(t_{i+1}) = x_\pi(t_i) + Ax_\pi(t_i)(t_{i+1} - t_i) + [Bx_\pi(t_i) + C]$$

$$[r(t_{i+1}) - r(t_i)] + \frac{1}{2} B\{[Bx_\pi(t_i) + C]$$

$$[r(t_{i+1}) - r(t_i)]\}[r(t_{i+1}) - r(t_i)] \; , \; i = 1,2,\ldots k \tag{3.8}$$

and by linear interpolation between $x_\pi(t_i)$ and $x_\pi(t_{i+1})$ for $t_i \leq t \leq t_{i+1}$. It is now possible to prove convergence of x_π to x when $m(\Pi)$ goes to zero: more precisely we have [16]:

Theorem 6

With assumptions a), b), c), d) of Theorem 5 we have:

$$\lim_{m(\Pi)\to 0} E\left[\| x(t) - x_\pi(t) \|^2\right] = 0 \qquad \text{uniformly in } [0,T]$$

* * *

The maximum likelihood method approach

Introducing the augmented state vector z defined in (1.1), from the stochastic model (3.1) and from (2.2), it is possible to get a model of the following form:

$$z(t) = \hat{z} + \int_0^t h[z(\tau)]d\tau + \int_0^t H_1[z(\tau)]dr(\tau) + \tag{3.10}$$

$$+ \frac{1}{2}\int_0^t H_2[z(\tau)]dr(\tau)dr(\tau)$$

$$y(t) = Gz(t) \tag{2.5'}$$

where h, H_1, H_2 are operators from R^m into (respectively) R^m, $R^m \times R^p$, $R^m \times R^p \times R^p$, and whose structure can be easily deduced from (3.1), (2.2).

The solution of (3.10) obviously has the same properties of existence and uniquenes enjoyed by the solution of (3.1).

Now we can deal with the problem of identification of the system (3.10), i.e. of estimation of $\hat{z}$ given the input and the output on $[0,T]$. Here and in the following we assume, for simplicity's sake that $\hat{z}$ is a deterministic unknown vector; we also assume that the realization of the input process r can be observed with no error (*). The observation of the output y, on the other hand, is corrupted by an additive white gaussian noise, with zero mean and unit covariance matrix. Actually, for greater precision, in the following we shall refer to the Wiener process and consequently deal not with the output process but with the process:

$$v(t) = \int_0^t Gz(\tau)d\tau + w(t) \;, \qquad t \in [0,T] \tag{3.11}$$

where W is a q-dimension Wiener process, with respect to

(*) The last hypothesis is introduced in order to avoid solving a nonlinear filtering problem which would otherwise be implied in the computation of the likelihood functional [17].

the family $\{F_t\}$.

Since both w and v are continuous stochastic processes, they induce a probability measure, respectively μ_w and μ_v, on the space $C[0,T]$ of q-dimensional continuous functions on $[0,T]$. Let us define the likelihood functional as the Radon-Nikodym derivative $d\mu_v/d\mu_w$ of μ_v respect to μ_w; the maximum likelihood estimate of $\hat{z}$ consists in that value of $\hat{z}$ which maximizes $d\mu_v/d\mu_w$ in the observed realization v.

Two problems now arise:

a) existence of the derivative $d\mu_v/d\mu_w$

b) computation of the likelihood functional for a given v.

As far as a) is concerned, we know [18] that $d\mu_v/d\mu_w$ exists if and only if μ_v is absolutely continuous with respect to μ_w. In this connection, we can prove the following theorem [16] which is based on a well-known theorem of Girsanov [19]:

Theorem 7

With assumptions a), b) of Theorem 5, if r is given by:

$$r(t) = Kw(t) + \int_0^t u(\tau)d\tau \qquad t \in [0,T] \tag{3.12}$$

where K is a p×q matrix, and u is measurable with $u(t) \in U$, $\forall\, t \in [0,T]$ for a compact set $U \subset R^p$, then the measure μ_v is equivalent and in particular absolutely continuous with respect to the measure μ_w and:

$$\frac{d\mu_v}{d\mu_w}(v) = \exp\left[\int_0^T z^*(\tau)G^*dv(\tau) - \frac{1}{2}\int_0^T z^*(\tau)G^*Gz(\tau)d\tau\right] \tag{3.13}$$

where the first integral on the RHS is a stochastic integral and z is the solution of (3.10). * * *

As far as b) is concerned, we note that the stochastic integral in (3.13) can be approximated, within any given accuracy, by a suitable finite sum [15]; the same obviously holds for the other (Lebesgue) integral. Thus, the difficulty reduces to the computation of z, i.e. to the solution of (3.10). However, we have shown that such a solution can be computed, within any given accuracy, by the procedure indicated in Theorem 6, for any given $\hat{z}$. Hence, the functional (3.13) is actually dependent only on $\hat{z}$, and the maximum likelihood estimate for the latter can be found by maximizing (3.13) with respect to $\hat{z}$ by any suitable numerical procedure.

References

[1] R.E. RINK and R.R. MOHLER,"Controllability and Optimal Control of Bilinear Systems",University of New Mexico, Bureau of Eng. Res., Tech. Rep. EE 143, June 1967.

[2] R.R. MOHLER, "Natural Bilinear Control Processes", IEEE Trans. on S.S.C., vol. SSC 6, n.3, 1970.

[3] R.R. MOHLER and R.E. RINK, "Reachable Zones for Equicontinuous Bilinear Control Processes", Int. J. of Control, vol.14 n.2, 1971.

[4] C. BRUNI, G. DI PILLO and G. KOCH, "On the Mathematical Models of Bilinear Systems", Ricerche di Automatica, vol. 2 n.1, 1971.

[5] E.A. CODDINGTON and N. LEVISON, "Theory of Ordinary Differential equations", Mc Graw Hill, 1955.

[6] C. BRUNI and G. DI PILLO, "On the Solution of Nonlinear Multipoint Boundary Value Problems by Newton's Method", Rapp. Ist. Automatica, Università di Roma, to appear.

[7] L.V. KANTOROVICH and G.P. AKILOV, "Functional Analysis in Normed Spaces", Mc Millan, 1964.

[8] P.L. FALB and J.L. DE JONG, "Some Successive Approximation Methods in Control and Oscillation Theory", Academic Press, 1969.

[9] L.B. RALL, "Computational Solution of Nonlinear Operator Equations", J. Wiley, 1969.

[10] R.E. KALABA, "On Nonlinear Differential Equations, the Maximum Operation and Monotone Convergence", J. Math. Mech., vol.8 n.4, 1959.

[11] R.E. BELLMAN and R.E. KALABA, "Quasilinearization and Nonlinear Boundary Value Problems", Elsevier, 1965.

[12] J.L. DOOB, "Stochastic Processes", J. Wiley, 1953.

[13] E.J. Mc SHANE, "Toward a Stochastic Calculus, I*", Proceedings National Acad. of Sciences, vol. 63, pag.275, 1969.

[14] E.J. Mc SHANE, "Toward a Stochastic Calculus, II*", Proceedings National Acad. of Sciences, vol.63, pag. 1084, 1969.

[15] E.J. Mc SHANE, "Stochastic Integrals and Stochastic Functional Equations", SIAM J. On Appl. Math. vol.17, n.2, 1969.

[16] G. KOCH, "Stochastic Bilinear Systems: Modeling and Identification", Rapp. Ist. Automatica, Università di Roma, to appear.

[17] T. KAILATH, "A General Likelihood - Ratio Formula for Random Signals in Gaussian Noise", IEEE Trans. on Information Theory, vol. IT 15 n.3, 1969.

[18] W. RUDIN, "Real and Complex Analysis", Mc Graw Hill, 1966.

[19] I.V. GIRSANOV, "On Transforming a Certain Class of Stochastic Processes by Absolutely Continuous Substitution of Measures", Theory of Probability and Its Applications, vol. 5 n.3, 1960.

ON THE ALGEBRAIC STRUCTURE OF BILINEAR SYSTEMS

Roger W. Brockett

Harvard University
Cambridge, Massachusetts

Abstract

In this paper we will show that input-output models of the form

$$\dot{x}(t) = (A + \sum_{i=1}^{m} u_i(t)B_i)x(t) \; ; \; y(t) = Cx(t)$$

are capable of representing a wide variety of highly nonlinear phenomena of practical importance and, at the same time, represent a class of systems for which a fairly detailed structural analysis can be made. The techniques needed are somewhat more algebraic than those which suffice for ordinary linear system theory. In particular, the theory of Lie algebras and their representations play a basic role.

Preliminary Ideas

Following a general theme in the mathematical theory of model building, our concern here is with the relationship between external (often emperical) descriptions of dynamic system, and internal (for us a description in terms of differential equations) descriptions of the model. We refer to the latter as a realization of an input-output system. The system itself is thought of as a collection of input-output pairs.

*This work was supported in part by the U.S. Office of Naval Research under the Joint Services Electronics Program by Contract N00014-67-A-0298-0006 and by the National Aeronautics and Space Administration under Grant NGR 22-007-172.

We want to describe a theory which is general enough to treat systems of the form

$$\dot{x}(t) = (A+\sum_{i=1}^{m} u_i(t)B_i)x(t)+\sum_{i=1}^{m} u_i(t)b_i \quad ; \quad y(t) = c[x(t)]$$

where A and B_i are square matrices, the b_i are column vectors and c[x] is a finite power series. This departure from linear systems, i.e. systems for which the $u_i(t)B_i$ terms are absent, and c is linear, is justified on the grounds that a number of practical control problems can only be modelled successfully if the multiplicative control and output nonlinearity are present. The reasons for this are explained, in part, by the observation that for problems where the state must satisfy bilinear or multilinear constraints, linear systems are necessarily uncontrollable, and hence degenerate, from an input-output point of view.

The theory of input-output models of the type

$$\dot{x}(t) = Ax(t) + Bu(t) \quad ; \quad y(t) = Cx(t)$$

draws on the theory of linear algebra. The ideas used all seem to center around subspaces which are invariant under A. The idea that all systems can be decomposed into a series-parallel innerconnection of real first and second order systems is of great theoretical and practical importance.

For bilinear models the tools of linear algebra are no longer enough. There is a simple explanation of this fact. In order to decompose the system equations as completely as possible, it is necessary to develop canonical forms for a set of matrices which admit both linear operations and a type of multiplication. A form which is convenient relative to the vector space structure of the set of matrices typically is not well behaved relative to the multiplicative structure and conversely. To sort this all out requires more than just linear algebra. For reasons having to do with controllability, the useful multiplication rule is [A,B] = AB-BA. The study of bilinear systems is intimately connected, therefore with the study of sets of matrices which are closed under vector space operation and also the above multiplication. These objects form Lie algebras and if we are to make reasonable progress in understanding bilinear systems, this theory cannot be avoided.

Examples leading to bilinear constraints include those where energy is to be conserved. If x must satisfy

$$x'Qx = 1$$

then we may model a controlled system by

$$\dot{x}(t) = (A + \sum_{i=1}^{m} u_i(t)B_i)x(t)$$

where $QA + A'Q = 0$ and $QB_i + B_i'Q = 0$.

Higher order constraints can also be accommodated. Let $V_1, V_2, \ldots, V_r$ and W be vector spaces over the same field. A map

$$\phi: V_1 \times V_2 \times \ldots \times V_r \to W$$

is called multilinear if it satisfies, for all α and β in the field and all $i = 1,2,\ldots r$.

$$\phi(v_1,v_2,\ldots,\alpha v_i+\beta v_i',\ldots,v_{r-1},v_r)$$
$$= \alpha\phi(v_1,v_2,\ldots,v_i,\ldots,v_{r-1},v_r)+\beta\phi(v_1,v_2,\ldots,v_i',\ldots,v_{r-1},v_r)$$

Given a multilinear form $\phi : \mathcal{R}^n \times \mathcal{R}^n \times \ldots \times \mathcal{R}^n \to \mathcal{R}$, suppose the constraint to be satisfied by x is

$$\phi(x,x,\ldots,x) = 1$$

Let the equations of motion be

$$\dot{x}(t) = (A + \sum_{i=1}^{m} u_i(t)B_i)x(t)$$

This imposes the conditions on A and B_i

$$L(Ax,x,\ldots x)+L(x,Ax,\ldots x)+\ldots+L(x,x,\ldots Ax) = 0$$

$$L(B_i x,x,\ldots x) + L(x,B_i x,\ldots x)+\ldots+L(x,x,\ldots B_i x) = 0$$

Specific instances which require both the additive and the multiplicative terms have been given in the literature [1]. One large class of problems of this type arise in the study of switched electrical networks, examples of which appear in [2] and [3]. The bilinear form is of basic importance in certain problems having a geometrical component due to the Frenet-Serret formulas for curves in a 3-dimensional space.

The Basic Bilinear Model

We want to show that a large class of input-output models can be reduced to the form

$$\dot{x}(t) = (A+\sum_{i=1}^{m} u_i(t)B_i)x(t) \;;\; y(t) = Cx(t) \qquad \text{(I)}$$

where x is an n-tuple, y is a q-tuple and A, $\{B_i\}$ and C are matrices of appropriate dimensions.

We begin with a simple observation. (Compare with [2] section 7 and [3] section 4.)

Theorem 1: Any input-output map which can be realized by a set of equations of the form

$$\dot{x}(t) = (A+\sum_{i=1}^{m} u_i(t)B_i)x(t)+\sum_{i=1}^{m} u_i(t)b_i \quad ; \quad y(t) = Cx(t)$$

can be realized by a set of equations of the form

$$\dot{z}(t) = (F+\sum_{i=1}^{m} u_i(t)G_i)z(t) \;;\; y(t) = Hz(t) \qquad \text{(I')}$$

Proof: Let F and G_i be defined by adding a single extra row and column to A, and B_i respectively

$$F = \begin{bmatrix} 0 & 0 \\ 0 & A \end{bmatrix} \qquad G_i = \begin{bmatrix} 0 & 0 \\ b_i & B_i \end{bmatrix}$$

Let z and H be given by

$$z = \begin{bmatrix} 1 \\ x \end{bmatrix} \qquad H = [0,\; C]$$

It is immediate that the z-system defines the same input-output map as the x-system. □

The second result is a little more involved. It shows that nonlinear output maps can be reduced to linear forms provided they are of the finite power series type. This is the kind of result that has no counter part in linear theory and points out the great flexibility inherit in the bilinear model. The basis for the result is the observation (which goes all the way back to the thesis of A.M. Liapunov) is that if x satisfies a linear equation then x(t)x'(t) satisfies one also. Thus in our case if x satisfies

(I) then (prime denotes transpose)

$$\frac{d}{dt} x(t)x'(t)=(A+\sum_{i=1}^{m} u_i(t)B_i)x(t)x'(t)+x(t)x'(t)(A+\sum_{i=1}^{m} u_i(t)B)'$$

which is an equation of the bilinear type! That is, there exist matrices $A^{[2]}$ and $B_i^{[2]}$ such that $z=(x_1^2,x_1x_2,x_1x_3,\ldots x_2^2,x_2x_3,\ldots,x_n^2)$ satisfies

$$\dot{z}(t) = (A^{[2]} + \sum_{i=1}^{m} u_i(t)B_i^{[2]})z(t)$$

Of course $A^{[2]}$ and $B_i^{[2]}$ are derived from A and B_i, respectively. One can be more explicit using Kronecker product relationships and the theory of symmetric tensors [4]. The same is true not only for $\{x_ix_j\}$ but also $\{x_ix_jx_k\}$ etc. as is easily verified. Thus associated with each bilinear equation is a countable collection of bilinear systems. The mth entry in this collection being the bilinear equation for the mth-degree forms in x. It can be taken to be of dimension equal to the number of linearly independent m-forms in n variables, i.e. $n(n+1)\ldots(n+m-1)/2.3\cdot\ldots m$. We indicate the vector consisting of these forms (ordered lexographically, for the sake of definiteness) by $x^{[m]}$.

Theorem 2: Any input-output map which can be realized in the form

$$\dot{x}(t)=(A+\sum_{i=1}^{m} u_i(t)B_i)x(t) \; ; \; y(t)=\sum_{p=1}^{q} L_p(x(t),x(t),\ldots,x(t))$$

where L_p is a p-linear map can be realized in the form

$$\dot{z}(t) = (F+\sum_{i=1}^{m} u_i(t)G_i)z(t) \; ; \; y(t) = Hz(t)$$

Proof: It is clear from the previous remarks that if x satisfies a bilinear state equation then so does $x^{[m]}$. Thus we can write an equation of the form

$$\dot{z}(t) = [\tilde{A} + \sum_{i=1}^{m} u_i(t)\tilde{B}_i]z(t)$$

where z is defined by

$$z' = (x,x^{[2]},\ldots x^{[q]})$$

and $[\tilde{A} + \sum_{i=1}^{m} u_i(t)\tilde{B}_i]$ is given by

$$\begin{bmatrix} A+u_i(t)B_i & 0 & \cdots & 0 \\ 0 & A^{[2]}+u_i(t)B_i^{[2]} & \cdots & 0 \\ \cdot & \cdot & \cdot & \cdot \\ 0 & 0 & & A^{[q]}+u_i(t)B_i^{[q]} \end{bmatrix}$$

Now y is a linear combination of the components of z since it is multilinear in the components of x. □

<u>Example</u>: The reader may verify that the input-output system defined by

$$\ddot{x} = u \quad ; \quad y = x^2$$

is represented by the bilinear system

$$\frac{d}{dt}\begin{bmatrix} 1 \\ x \\ \dot{x} \\ x^2 \\ \dot{x}x \\ \dot{x}^2 \end{bmatrix} = \begin{bmatrix} 0 & 0 & 0 & 0 & 0 & 0 \\ 0 & 0 & 1 & 0 & 0 & 0 \\ u & 0 & 0 & 0 & 0 & 0 \\ 0 & 0 & 0 & 0 & 2 & 0 \\ 0 & u & 0 & 0 & 0 & 1 \\ 0 & 0 & 2u & 0 & 0 & 0 \end{bmatrix} \begin{bmatrix} 1 \\ x \\ \dot{x} \\ x^2 \\ \dot{x}x \\ \dot{x}^2 \end{bmatrix}$$

$$y = [0 \quad 0 \quad 0 \quad 1 \quad 0 \quad 0]x$$

System Interconnection

We say that two bilinear systems (I) and (I') are <u>interconnected in parallel</u> to get the single system if we simply add their outputs. That is, the equations for the parallel interconnection are

$$\dot{x}(t)=(A+\sum_{i=1}^{n} u_i(t)B_i)x(t)$$

$$; \ y(t) = Cx(t)+Hz(t)$$

$$\dot{z}(t)=(F+\sum_{i=1}^{n} u_i(t)G_i)z(t)$$

Clearly this is defined only if the dimensionality of the

input spaces of I and I' are the same and the dimensionality of the output spaces of the two systems are the same.

We say that two bilinear systems are interconnected in series with (I') following (I) if the input to (I') is equated to the output of (I) the equations for the series interconnection are

$$\dot{x}(t) = (A + \sum_{i=1}^{m} u_i(t)B_i)x(t)$$

$$; \; y(t) = Hz(t)$$

$$\dot{z}(t) = (F + \sum_{i=1}^{m} (Cx)_i(t)G_i)z(t)$$

Clearly a series connection is possible if the dimension of the output of the first system equals the dimension of the input of the second.

Remark: If the parallel interconnection of two input-output systems having bilinear realizations is defined then the system which results from parallel interconnection has a bilinear realization. If the series connection of a system having a bilinear realization followed by a system having a linear realization is defined, then the system which results from series interconnection has a bilinear realization.

We have not been able to determine if the class of bilinear realizations is closed under series interconnection.

The Canonical Form

The existance of the Jordan normal form for a linear map of $\mathscr{R}^n$ into $\mathscr{R}^n$ gives rise to the "diagonal" or "partial fraction" realization for linear systems. This is important because in certain senses the Jordan form displays the maximum degree decoupling which is possible. We want to describe the analogous situation for bilinear systems. As might be expected, the results cannot be based on the tools of linear algebra alone.

In view of the results of section 2, we are content to consider hence forth systems which have realizations in the form of equation (I).

We call two realizations

$$\dot{x}(t) = (A+\sum_{i=1}^{m} u_i(t)B_i x(t) \quad ; \quad y(t) = Cx(t) \qquad \text{(I)}$$

and

$$\dot{z}(t) = (F+\sum_{i=1}^{m} u_i(t)G)x(t) \quad ; \quad y(t) = Hz(t) \qquad \text{(I')}$$

equivalent if there exists a nonsingular P such that $PAP^{-1} =$ F and $PB_iP^{-1} = G_i$ and $CP^{-1} = H$.

We call a realization in the form (I) irreducible if there is no nonsingular P such that

$$PAP^{-1} = \begin{bmatrix} \tilde{A}_{11} & 0 \\ \tilde{A}_{21} & \tilde{A}_{22} \end{bmatrix} \quad PB_iP^{-1} = \begin{bmatrix} \tilde{B}^i_{11} & 0 \\ \tilde{B}^i_{21} & \tilde{B}_{22} \end{bmatrix}$$

where $\tilde{A}_{11}$ and $\tilde{B}^i_{11}$ are square matrices, all of the same dimension. That is, for no choice of basis is the realization in block triangular form. Otherwise we call it reducible. A reducible realization is said to be completely reducible if it can be put in block diagonal form (as opposed to block triangular form) with each block being irreducible. A realization of the form (I) said to be equivalent to a triangular realization if there exists a nonsingular P (possibly complex) such that PAP^{-1} and PB_iP^{-1} are lower triangular. (Including the possibility of nonzero elements on the diagonal.) We call it strictly triangular if there exists P such that PAP^{-1} and PB_iP^{-1} are strictly lower triangular. (No nonzero elements on the diagonal.)

If a system is reducible then there are nontrivial invariant subspaces for the collection of matrices $\{A,B_i\}$. Let V_1 be one of these which is of smallest (positive) dimension. (There may be many, pick any one.) Let V_2 be a smallest invariant subspace properly containing V_1. Let V_3 be a smallest invariant subspace properly containing V_2, etc. Let $n_i = \dim V_i$. Pick a basis such that the first n_1 elements span the space V_1, the first n_2 elements span V_2, etc. Relative to this basis the matrices A and B_i take the block triangular form

$$A = \begin{bmatrix} A_{11} & 0 & 0 & \cdots \\ A_{12} & A_{22} & 0 & \cdots \\ A_{13} & A_{23} & A_{33} & \cdots \\ \cdot & \cdot & \cdot & \cdot \end{bmatrix} \quad B_i = \begin{bmatrix} B^i_{11} & 0 & 0 & \cdots \\ B^i_{12} & B^i_{22} & 0 & \cdots \\ B^i_{13} & B^i_{23} & B^i_{33} & \cdots \\ \cdot & \cdot & \cdot & \cdot \end{bmatrix}$$

Each of the collection of block diagonals $\{A_{kk}, B^i_{kk}\}$ are irreducible and the Jordan-Hölder Theorem insures that these representations are unique in that regardless of how the invariant subspaces are chosen, the construction will lead to an equivalent collection of irreducible diagonal blocks. (They may occur in a different order depending on the choice of subspace, of course.) We collect these observations in a theorem. (See, e.g. Samelson [4] page 12 for a sketch of a proof.)

Theorem 3: Every bilinear realization (I) is equivalent to one in which the A and B_i matrices are in block triangular form with the diagonal blocks being irreducible. Moreover if (A,B_i,C) and (F,G_i,H) are two equivalent realizations in block triangular form with irreducible blocks on the diagonal then there is a permutation π and nonsingular matrices P_k such that the diagonal blocks are related by

$$P_k A_{kk} P_k^{-1} = F_{\pi(k)\pi(k)} \quad ; \quad P_k B_{kk} P_k^{-1} = G_{\pi(k)\pi(k)}$$

We will say that an input-output system displayed according to the above recipe is in a reduced form.

Controllability

A detailed study of the controllability properties of bilinear and even more general systems, has been made in the recent literature. References [5] - [8] contain many interesting results. For our present purposes section 7 of [2] and section 6 of [8] are relevant.

In reference [2] it is shown that if A is zero, or if a certain commutation condition is satisfied, then the reachable set for

$$\dot{x}(t) = (A + \sum_{i=1}^{m} u_i(t)B_i)x(t) \; ; \; y(t) = Cx(t) \; ; \; x(0) = x_o \quad \text{(I)}$$

is easily computed. However Jurdjevic and Sussmann [8] have shown that the reachable set for (I) contains an open subset of the set reachable for

$$\dot{x}(t) = (v(t)A+ \sum_{i=1}^{m} u_i(t)B_i)x(t) \; ; \; y(t)=Cx(t) \; ; \; x(0) = x_o \qquad \text{(II)}$$

From this fact it is easy to show that the reachable set for (I) is confined to a subspace if and only if the reachable set for (II) is confined to the same subspace. We omit the details but make explicit use of this result below.

Theorem 4: The reachable set for (I) is confined to a subspace if and only if there exists a nonsingular P such that

$$PAP^{-1} = \begin{bmatrix} \tilde{A}_{11} & 0 \\ \tilde{A}_{21} & \tilde{A}_{22} \end{bmatrix}$$

$$PB_iP^{-1} = \begin{bmatrix} \tilde{B}^i_{11} & 0 \\ \tilde{B}^i_{21} & \tilde{B}^i_{22} \end{bmatrix}$$

$$P^{-1}x_o = \begin{bmatrix} 0 \\ \tilde{x}_o \end{bmatrix}$$

where the 0 blocks are all of the same dimension.

Proof: If there exists such a P then clearly the reachable set is confined to the subspace consisting of those vectors whose upper portion is zero.

Suppose the reachable set of (I) is confined to a subspace. Then by our remarks above the reachable set for (II) is confined to a subspace. But from the results of section 7 of [2] we see that this implies that A and B_i can be simultaneously block triangularized. □

Remark: Notice that the set of matrices $\{A,B_i\}$ can be simultaneously triangularized if and only if one can simultaneously triangularize the larger set obtained from $\{A,B_i\}$ by adjoining all linear combination products of any two elements, products of products, etc. More precisely, we define $\{A,B_i\}_{AA}$ to be the smallest vector space of matrices which contains $\{A,B_i\}$ and is closed under multiplication by elements of $\{A,B_i\}$. This larger set is called the associative algebra.

generated by $\{A,B_i\}$. The condition of Theorem 4 can be stated as requiring that x_o should not belong to any subset of $\mathscr{R}^n$ which is invariant with respect to multiplication by elements of the associative algebra. This statement is close to the familiar $(B,AB,A^2B,\ldots)$ test for controllability.

Theorem 5: Any input-output map which can be realized by a bilinear system can be realized by one for which the reachable set is not confined to a linear subspace.

Proof: Use Theorem 4. If the reachable set is confined to a subspace find the P which effects the decomposition for Theorem 4. Delete the top block, then the input-output map is the same but the state is not confined to a linear subspace. □

Observability

We will say that two starting states, x_o and x_1 of the system

$$\dot{x}(t) = (A+\sum_{i=1}^{m} u_i(t)B_i)x(t) \quad ; \quad y(t) = Cx(t) \tag{I}$$

are indistinguishable if for all inputs u, the response y is the same. This follows our approach in [2] where more general output maps are considered. We start off the analog of Theorem 4.

Theorem 6: The system (I) has no indistinguishable states if and only if there exists no nonsingular P such that CP^{-1}, PAP^{-1}, and PB_iP^{-1} take the form

$$CP^{-1} = [C, 0]$$

$$PAP^{-1} = \begin{bmatrix} A_{11} & 0 \\ A_{21} & A_{22} \end{bmatrix} \quad ; \quad PB_iP^{-1} = \begin{bmatrix} B_{11}^i & 0 \\ B_{21}^i & B_{22}^i \end{bmatrix}$$

Proof: Clearly if such a P exists then the system is not observable since $x_o = (0,\tilde{x})$ implies $y = 0$.

On the other hand, if there exists two indistinguishable states then there is a hyperplane of indistinguishable states. Hence

$$C\Phi_{(A+\Sigma u_i B_i)}\mathscr{K} = 0$$

for some subspace $\mathcal{K}$. Let x be in $\mathcal{K}$ then x belongs to the kernel of C. Thus we may characterize $\mathcal{K}$ as the largest subspace of the kernel of C which is invariant under the action of* $\Phi_{(A+\Sigma u_i B_i)}$. If such a subspace exists then there exists a choice of basis such that $(A,\{B_i\},C)$ has the form indicated. □

The remark following Theorem 4 is relevant here as well.

We now give the observability version of Theorem 5.

Theorem 7: Any input-output map which can be realized by a bilinear system can be realized by one for which there are no indistinguishable states.

Proof: Use Theorem 6. If there are indistinguishable states then triangularize the system and delete the lower part of the systems. If the resulting system has indistinguishable states repeat the operation until there are no more indistinguishable states. □

Example: We can apply these results to a linear system with a linear or power law output. The n-dimensional scalar input, scalar output system

$$\dot{x} = Ax + bu \quad ; \quad y = (cx)^2 \quad ; \quad x(0) = 0$$

takes the form

$$\frac{d}{dt}\begin{bmatrix} 1 \\ x \\ x^{[2]} \end{bmatrix} = \begin{bmatrix} 0 & 0 & 0 \\ ub & A & 0 \\ 0 & uB & A^{[2]} \end{bmatrix}\begin{bmatrix} 1 \\ x \\ x^{[2]} \end{bmatrix} \quad ; \qquad y = [0,\ 0,\ \tilde{C}]\begin{bmatrix} 1 \\ x \\ x^{[2]} \end{bmatrix} \qquad (*)$$

Now if $(b, Ab,\ldots,A^{n-1}b)$ is of rank $n = \dim x$, then there is no vector space which contains the reachable set for the realization (*). The observability criterion can be applied to show that this system has no distinct indistinguishable states if $c; cA;\ldots cA^{n-1})$ is of rank n and cA^ib is nonzero for some i.

*Here and above Φ with a subscript refers to a transition matrix associated with a linear system. See [9] section 4.

Equivalent Realizations

The state space isomorphism theorems for automata and linear systems are well known and of basic importance in these fields. Recently theorems of this type have appeared in other settings, for example [2] and [10]. Here we want to describe such a result for bilinear systems.

In this section we show that any two bilinear realizations of the same input-output may differ at most by a change of basis provided some natural minimality conditions are satisfied.

Let us agree to call x_o an <u>equilibrium</u> state of the bilinear system

$$\dot{x}(t) = (A+\sum_{i=1}^{m} u_i(t)B_i)x(t) \;;\; y(t) = Cx(t) \qquad \text{(I)}$$

if Ax_o vanishes. This is the same as asking that x_o be an equilibrium solution of the differential equation which results when all the u_i are set to zero.

<u>Theorem</u> : Suppose that we are given two realizations of the same input-output map

$$\dot{x}(t) = (A+\sum_{i=1}^{m} u_i(t)B_i)x(t);y(t) = Cx(t) \;;\; x(0) = x_o$$

$$\dot{z}(t) = (F+\sum_{i=1}^{m} u_i(t)G_i)z(t);y(t) = Hz(t) \;;\; z(0) = z_o$$

Let x_o and z_o be equilibrium states. Suppose that both systems are observable in that any two starting states can be distinguished for a suitable choice of u and suppose that the systems are controllable in that the reachable set from x_o or z_o is not confined to any proper linear subspace. Then the two realizations are equivalent.

<u>Proof:</u> Let the z-system be of dimension n. Without loss of generality we can assume the x-system is of dimension less than or equal to n. Let $u^1,u^2,\ldots,u^n$ be controls which are defined over the intervals $[0,t_1]$, $[0,t_2],\ldots[0,t_n]$ which result in z-trajectories $z^1,z^2,\ldots,z^n$. Let t_* be the largest of the t's and define $\tilde{u}^1,\tilde{u}^2,\ldots,\tilde{u}^n$ on $[0,t_*]$ by shifting the u^i to the latter portion of the interval and filling in

on the first portion with 0.

$$\tilde{u}^i(t) = \begin{cases} 0 & 0 \leq t \leq t_* - t_i \\ u^i(t-t_*+t_i) & t_*-t_i \leq t \leq t_* \end{cases}$$

Let $\tilde{z}^i$ be the resulting trajectory in the z system. As a result of the assumption that z_o is an equilibrium state

$$\tilde{z}^i(t) = \begin{cases} z_o & 0 \leq t \leq t_*-t_i \\ \tilde{z}^i(t-t_*+t_i) \; ; & t_*-t_i \leq t \leq t_* \end{cases}$$

Let x^i be the trajectory which the x-systems generates under the control u^i. Because both systems generate the same input-output map we have

$$C\Phi_{(A+\Sigma u_i B_i)}(\tilde{x}^1(t_*),\tilde{x}^2(t_*),\ldots\tilde{x}^n(t_*))$$

$$= H\Phi_{(F+\Sigma u_i G_i}(\tilde{z}^1(t_*),\tilde{z}^2(t_*),\ldots\tilde{z}^n(t_*))$$

where $\Phi_{(A+\Sigma u_i B_i)}$ and $\Phi_{(F+\Sigma u_i G_i)}$ are the transition matrices which result from an arbitrary control u.

Now the matrix $Z = (\tilde{z}^1(t_*),\tilde{z}^2(t_*),\ldots,\tilde{z}^n(t_*))$ is nonsingular by construction. If x-system is not of the same dimension as the z-system, or if the matrix $X = (\tilde{x}^1(t_*),$ $\tilde{x}^2(t_*),\ldots,\tilde{x}^n(t_*))$ is singular then there exists a nonzero vector η such that $X\eta = 0$.

$$C\Phi_{(A+\Sigma u_i B_i)}X\eta = H\Phi_{(F+\Sigma u_i G_i)}Z\eta = 0$$

Thus $Z\eta$ is a starting state for the Z-system which is nonzero but equivalent to 0. This violates the observability hypothesis. Thus X must be a square matrix which is nonsingular.

Since we have for all u

$$C\Phi_{(A+\Sigma u_i B_i)}X = H\Phi_{(F+\Sigma u_i G_i)}Z$$

and since I is certainly a possible transition matrix

$$CXZ^{-1} = H$$

Moreover, since no two states give rise to the same input-output map, the equality

$$C\Phi_{(A+\Sigma u_i B_i)} = CXZ^{-1}\Phi_{(F+\Sigma u_i G_i)}ZX^{-1}$$

implies

$$\Phi_{(A+\Sigma u_i B_i)} = XZ^{-1}\Phi_{(F+\Sigma u_i G_i)}ZX^{-1}$$

From this it follows that for $P = XZ^{-1}$

$$A = PFP^{-1}$$

$$B_i = PG_iP^{-1}$$

and from above

$$C = HP^{-1}$$

□

This result can also be used to establish isomorphism theorems for realizations in inhomogeneous form. That is, two realizations of the form

$$\dot{x}(t)=(A+\Sigma u_i(t)B_i)x(t)+\sum_{i=1}^{m} u_i(t)b_i \qquad y(t) = Cx(t)$$

can be shown to differ only by a choice of basis provided the appropriate minimality conditions are satisfied.

We point out that in actually determining equivalent realizations for systems and in the classification of systems, the results available in the study of Lie algebras (e.g. [4]) are of fundamental importance. Some recent work relating Lie algebras and system theoretic ideas is reported in [11].

Conclusions

In this paper we have shown that a particular bilinear model is both quite general and easy to work with. Building on previous results we have shown how to get a basic structure theory. There are many more specific problems which can be examined using these tools. Some of these are under investigation and will be reported on soon.

References

1. R.E. Rink and R.R. Mohler, "Completely Controllable Bi-Linear Systems," SIAM J. Control, Vol. 6, No. 3, 1968.
2. R.W. Brockett, "System Theory on Group Manifolds and Coset Spaces," SIAM J. on Control, May 1972.
3. R.W. Brockett, "Algebraic Decomposition Methods for Non-linear Systems," in System Structure, IEEE Special Publication No. 71C61-CSS, August 1971 (S. Morse, Editor).
4. H. Samelson, Notes on Lie Algebras, Van Nostrand Co. 1969.
5. C. Lobry, "Controllabilite des systems non linearies," SIAM J. Control, Vol. 8, (1970), 573-605.
6. G.W. Haynes and H. Hermes, "Nonlinear Controllability via Lie Theory," SIAM J. Control, Vol. 8, (1970).
7. H. Sussmann and V. Jurdjevic, "Controllability of Non-linear Systems," J. of Differential Equations, to appear.
8. V. Jurdjevic and H.J. Sussmann, "Control Systems on Lie Groups," J. Differential Equations, to appear.
9. R.W. Brockett, Finite Dimensional Linear Systems, John Wiley and Sons, N.Y. 1970.
10. R.W. Brockett and A. Willsky, "Finite State Group Homomorphic Sequential Systems," IEEE Trans. on Automatic Control, August 1972.
11. R.W. Brockett and A. Rahimi, "Lie Algebras and Linear Differential Equations," in Ordinary Differential Equations, Academic Press, 1972. (L. Weiss, Editor).

BILINEAR STRUCTURES AND MAN

R. R. MOHLER *

Department of Electrical and Computer Engineering
Oregon State University
Corvallis, Oregon 97331

Introduction

The purpose of this paper is to summarize past research on controllability and optimal control of finite state bilinear systems, (i.e., systems linear in control and in state but not linear in both), and then to discuss recent results for these variable structures in living systems.

Though linear system models frequently are utilized to approximate real dynamical processes for convenience, their rigid structure or inability to adapt to inputs constrains their usefulness for many natural processes. Such limitations are particularly apparent in socio-economics and in life science. The single population equation for biological species is perhaps the simplest example. The rate of change of population is:

$$\frac{dx}{dt} = ux \qquad (1)$$

where u, birth rate minus death rate, may be considered a control variable, and the state x is population.

Despite the simplicity of (1) with variables separable, it is not uncommon to assume constant control over short time intervals or to predict future control values based on that of the past for the species of interest and on more extended histories of other classes. For example,

*This research is supported by the National Science Foundation, Grant No. 33249.

past u(t) for Europe (already subsiding) as well as for Asia (still rising but near the peak) is used to predict future populations of Asia. Also, feedback models of various degrees of complication have been derived to describe various constraints built into the species to provide the necessary structure by means of nonlinear closed loop system (1).

So at the extremes, linear analysis and nonlinear analysis may be useful. But by analyzing the bilinear model (1) a better understanding of population control may be obtained.

While the biological population equation (1) is somewhat of a degenerate bilinear model it is not difficult to find similar bilinear population equations with added complications. For example, generation of cells by biological fission or of neutrons by nuclear fission may be modeled by more complicated bilinear equations. In the latter case, delayed neutrons are generated from several precursors or unstable fission products. For U-235 fissions, the state may be defined by the corresponding levels of neutrons and six such precursors. Similarly, the buildup of xenon and iodine poison in a nuclear reactor is described by a bilinear system. Such nuclear processes and optimal control of nuclear reactors are analyzed by Mohler and Shen [2].

Conductive and convective heat-transfer processes may frequently be described by bilinear models [2]. A bilinear model for thermal regulation of a human body is discussed below along with bilinear compartmental biological systems.

For generation of animal beef the bilinear model includes control by feed regulation. Socio-economic processes frequently are modeled in an awkward manner by linear or piecewise linear equations. Again, a bilinear model arises in a natural manner similar to the population equation. For the economy of the Chinese Peoples Republic, an appropriate bilinear model utilizes constant control over periods of somewhat constant policy. From 1950 to 1957 her economic "time constant", u^{-1}, was about 5 years. This was followed by a period of even faster economic gains, (the "great leap"), but by over expansion and eventual recession in the early sixties. Again, there was a brief period of economic resurgence in the mid sixties similar to that just prior to 1957. Then a new policy instigated by the "cultural revolution" caused at least a temporary slow down. Of course, such a model is a gross simplification of the actual process, but it is doubtful that sufficient data is available to justify further refinement at this time even if it is desired.

The braking action so common to the automobile is approximated by a bilinear model. Here, the frictional force is nearly proportional to the product of the normal force applied to the brake (control variable) and the speed of the automobile (state variable). While dynamical structure is varied by the brake pedal, additive control is available by the accelerator or gas pedal. These bilinear models are examined in more detail by Mohler [3,4].

The bilinear systems studied here are defined by the general state equation of form $\frac{dx}{dt} = f(x,u)$ or

$$\frac{dx}{dt} = Ax + B(x,u) + CU, \qquad (2)$$

where $x \varepsilon R^n$ is the state vector, $u \varepsilon\ U \subset R^m$ is the control vector, A and C are nxn and nxm real matrices. B(x,u) is bilinear mapping from $R^n \times R^m$ into R^n. It is the product of state and control,

$$B(x,u) = \sum_{k=1}^{m} u_k B_k x \qquad (3)$$

where B_k is an nxn real constant matrix, which of course provides the variable structure to an otherwise jointly linear system. The bilinear state diagram is given by Fig. 1.

Also, it is assumed there that u(t) is at least piecewise continuous and that the admissible set U is compact, simply connected and contains the origin.

Controllability

Though the familiar rank test may be made to determine complete controllability of linear systems, such degree of controllability seldom occurs for practical linear control systems.

Consider the classic linear system (Eq. (2) without $B(x,u)$ with all eigenvalues of A in the left-hand complex plane and with $u \varepsilon U$. From the existence of a quadratic Liapunov function $V(x)$ for $\dot{x} = Ax$, it is obvious from the inner product,

$$\dot{x} \cdot \frac{\partial V}{\partial x} = \dot{V}(x) + Cu \cdot \frac{\partial V}{\partial x}, \tag{4}$$

that for $\|x\|$ sufficiently large, there exists a suffiently large M such that $\dot{x}$ is directed into the region bounded by $V(x) = M$. Hence, the asymtotically stable linear system is not completely controllable with bounded control.

The variable structure of the bilinear system (linear system with $B(x,u)$ added) allows it to be more controllable just as it frequently provides a more accurate model.

For example, consider the linear system:

$$\frac{dx_1}{dt} = x_2 \tag{5}$$

$$\frac{dx_2}{dt} = -2x_1 - x_2 + u \tag{6}$$

First, if $u(t)$ is not bounded it is obvious from the rank test that (6), is completely controllable. From the previous result, however, it is obvious that this stable linear system with bounded control would not be completely controllable. This system, however, would be locally controllable about the origin - a statement which is geometrically obvious in state space but which may also be shown by a theorem due to Lee and Markus [5]. Since controllability is really a property of portrait connectedness in state space, it is readily apparent that the combination of local controllability and proper manipulation of system eigenvalues with admissible controls may make the system completely controllable.

If $u \leq$ and if $(x_1 + x_2)u$ is added to the right hand side of (6), the linear system becomes bilinear and is completely controllable. For $u(t)$ at the extreme values of

+1 and -1, the state portrait includes a superposition of unstable and stable foci portraits with equilibrium states at (1,0) and (- 1/3, 0) respectively.

Rink and Mohler [6] show more precisely that (3) with $u \varepsilon U$ is completely controllable if:

1. There exist control values, u^+ and u^-, such that the real parts of the eigenvalues of the system matrix,

$$A + \sum_{k=1}^{m} u_k B_k,$$

are positive and negative, respectively, and such that the equilibrium states $x^e(u^+)$ and $x^e(u^-)$ are contained in a connected component of the equilibrium set;

2. For each x in the equilibrium set with equilibrium control $u^e(x)$, there exists a $v \varepsilon R^m$ such that Dv, EDv, ..., $E^{(n-1)}$ Dv are linearly independent, where $D = \frac{\partial f}{\partial u}(x, u^e(x))$, and

$$E = \frac{\partial f}{\partial x}(x, u^e(x)).$$

For canonical phase-variable systems with $x_1 = x$, $x_2 = \dot{x}$, ..., $x_n = x^{(n-1)}$, however, the second condition (local controllability) is satisfied if C is simply a nonzero matrix. Also, the first condition (connectedness) is satisfied if all the eigenvalues of the system matrix can be shifted across the imaginary axis, without passing through the origin, as u ranges continuously over a subset of U.

Optimal Control

The existence of optimal control policies for bilinear systems is readily established from existence theorems for optimal controls [7]. Then several computational procedures are available that have been developed for bilinear control processes. One such scheme to design optimal bilinear regulators is based on the maximum principle along with certain sufficiency conditions. In regions of state space where the maximum principle does not yield unique solutions a truncation procedure is used to form continuous cost surfaces such that optimality is assured. This work is discussed in detail by Mohler and Rink [8,9]. The method has provided an excellent suboptimal synthesis of a bilinear servomechanism.

It is apparent from the maximum principle that bang-bang processes are candidates for the optimal control of bilinear processes which have a performance index that is at least linear in control. For these problems a modified gradient algorithm called the switching-time-variation method, STVM, has been derived which generates a sequence of step-control processes to approximate successively an extremal control [7]. The STVM uses the linear control property of bilinear systems to yield a most efficient algorithm when compared to other techniques. Mohler and Moon [7] show the convergence of this sequence to a maximum principle solution.

Compartmental Living Systems

Discretization or compartmentation of complicated distributed processes for mathematical or geometrical convenience is quite common in practice. While there are many interesting results published on mathematical aspects of discretization and its use in optimal control, here anatomical compartmentation in living systems is the subject of primary interest.

A living cell might represent one single compartment of this sort. Still, the biochemical processes internal to the cell itself are not so simple. Enzymes must be present to activate necessary reactions which lead to the formation of proteins, nucleic acids and their precursors. These in turn provide cell growth.

Enzymes act as catalysts to provide the necessary degree of controllability. For biochemical cellular processes, reactions may be approximately described by the law of mass action just as for general chemical reactions under the control of catalysts. In this way, reaction rates are assumed to be proportional to reacting concentrations with bilinear control or catalyst concentrations [4,10]. While a single cell may have thousands of series and parallel reactions to this type taking place, only a few pacemakers are needed to model the cellular plant for most purposes.

Control by enzyme concentration is manipulated in the cell through the DNA.

Molecules of compounds formed from cellular biochemical processes or metabolism move in all directions in a random manner. There molecules diffuse naturally from a region of high concentration to a region of lower concentration with a rate nearly proportional to the product of separating membrane permeability and difference in concentrations. Again bilinear control is synthesized by enzymes which in this case manipulate membrane permeability.

In addition to such natural diffusion, however, material may be transported in the opposite direction (against the concentration gradient) by a phenomenon called active transport. The energy for active transport is derived from cellular metabolism. While the active transport process is quite nonlinear [11] it may be approximated by a product of linear control terms and linear concentration terms (possible state variables).

The same form of equations may be generated for large anatomical compartments within a living system just as for single cells. Such processes include fluid and electrolyte balance, kinetics of material injected or ingested into or excreted from the body and kinetics of metabolites in cell suspensions or tissues [4,12,13].

Though it is assumed implicitly in the discussions here that mass transfer is the process of concern, it should be realized that the concepts can be generalized to include generalized fluxes between compartments such as for regulation of body temperatures, carbon dioxide in the lungs and systemic pressures in the arteries [4,14]. It is interesting to note here that the bilinear control for these processes includes ventilation rate, vasomotor variation of thermal conductivity and vasomotor variation of resistance to blood flow.

The n compartments of such physiological systems are related by conservation equations of the form

$$\frac{dx_i}{dt} = \sum_{j=1}^{n}{}' \phi_{ij} - \sum_{k=1}^{n}{}' \phi_{ki} + \phi_{ia} - \phi_{ai} + p_i - d_i, \qquad (7)$$

$i = 1,\ldots,n,$

where x_i is the concentration of substance X in the ith compartment, ϕ_{ij} is the flux of X from the jth to the ith compartment, ϕ_{ia} (ϕ_{ai}) is the flux of X into (out of) the ith compartment from (to) the environment to a uniform concentration x_a of X, $p_i(d_i)$ is the rate of production (destruction) of X in ith compartment. And, primed sigma denotes deletion of the ith term. Compartmental volumes are assumed constant for convenience.

The x_i are regulated by homeostasis through the manipulation of fluxes, of production rates and of destruction rates of X within compartments.

Flux from jth compartment to ith compartment is described by

$$\phi_{ij} = \rho_{ij} x_j, \tag{8}$$

where i, j=1,...,n and ρ_{ij} is an exchange constant. It is obvious that control enters the system linearly through the net production of substance ($p_i - d_i$). If this net production is zero, the system is called "conservative" and only parametric control remains.

Tracer Experiments

For most physiological processes it is not at all convenient to decouple sybsystems, to measure all controls and to measure all state variables. Usually, it is not practical to add test signals to the physiological process that is to be modeled. Unfortunately, this precludes the use of many of the most convenient properties of bilinear systems. Such limitations are especially prevalent in human-body experiments. For many of these processes, compartmentation lends itself conveniently to modeling and identification by means of tracer experiments since such experiment may be conducted without having a noticeable effect on the regulated process.

Isotopic tracers are particularly convenient and usually harmless to identify compartmental substance concentrations. Usually it is assumed: (1) that the tracer is distributed uniformly throughout the substance X, (2) that labeled and unlabeled X behave identically, and (3) that the tracer does not alter the system dynamics. Obviously, this requires that the amount of tracer in the system is neglected.

Suppose tracer is inserted directly into p system compartments and that its behavior can be monitored in q compartments. Define an input matrix, P, with unity elements in each row corresponding to a compartment accessible for tracer insertion and with zero elements elsewhere. Each column of P has exactly one unity element and the rank of P is p. Also, define a q x n output matrix Q with zero elements except for unity in each column which corresponds to an observable compartment.

Then from (7) with specific tracer activity, $a \in R^n$, tagging X (by multiplications), with an added tracer insertion activity, $f_p \in R^p$ and with system at "equilibrium", experimental access is described by

$$\frac{da}{dt} = S\,a + X_d^{-1}\,p\,f_p \tag{9}$$

$$w = Q\,a, \tag{10}$$

where X_d is the diagonal matrix of elements $x_1,\ldots,x_n$ (assumed to be at "equilibrium"), f_p is the p-dimensional vector of inserted total activity fluxes, and therefore, $X_d^{-1}\,p\,f_p$ is the vector of inserted specific activity fluxes. W is the q-dimensional output vector of observed compartmental specific activities [4]. Also, $S = s_{ij}$ with

$$s_{ij} = \phi_{ij}/x_i \qquad j = 1,\ldots,n, \quad i \neq j$$

and

$$s_{ii} = -\frac{1}{x_i} \sum_{\ell=1}^{n} \phi_{i\ell}\,.$$

Now the rank tests for controllability and observability may be applied to the linear tracer dynamics to see if sufficient accessibility is available to identify the minimum realization of the system. Fig. 2 shows several examples where tracer is inserted/observed in various compartments.

Now, consider the experiment with tracer entering the system by naturally available flux routes, Assume that fluxes ϕ_{ia} with specific activities a_{ia}, i=1,...,n, originate from p´ separate external sources of labeled substance, each with its own tracer specific activity a_i', $i=1,\ldots,p'$. In this case

$$\frac{da}{dt} = S\,a + X_d^{-1}\,\Phi_{ad}\,P'\,a', \qquad (11)$$

and again

$$w = Q\,a, \qquad (12)$$

where Φ_{ad} is a diagonal matrix of elements $\phi_{ia} \ldots \phi_{na}$, a' is p - dimensional tracer source activity vector, and P' is the n x p´ input matrix with a single unity in each row, i, for which $a_{ia} \neq 0$. Since more than one compartment may receive substance from the same external source, P' may have more than a single unity in a given column. The rank test for complete observability is as before. If no two compartments receive tracer from the same source, then this matrix is the same as for the previous case with P' replacing P.

Commonly used tracer "soak up" and "wash out" experiments fit the latter category of "natural" tracer absorption of tracer monitored in a "soak up" experiment. In tracer "wash out" experiments, the system is put in a tracer-free environment so that wash out is observed. In either case, the present of all compartments can not be detected unless the rank test is satisfied.

Conclusions

As a consequence of their variable structure, bilinear systems are appropriate for modeling many natural phenomena in man and his environment. Just as nature seems to be predominantly adaptive, a variable-structure, such as that of bilinear control, frequently can offer better performance than can a linear system. The simple concepts of controllability and observability are found to be quite convenient to determine the necessary accessibility of tracer compartments for minimum realizations.

References

1. Pielou, E., An Introduction to Mathematical Ecology, Wiley-Inter Science, New York, 1969.

2. Mohler, R. R. and Shen, C. N., Optimal Control of Nuclear Reactors, Academic Press, New York 1970.

3. Mohler, R., "Natural Bilinear Control processes", IEEE Trans. Sys. Sci. and Cyb., SSC-6, 192-197 (1970).

4. Mohler, R. R., Bilinear Control Processes with Applications to Engineering, Ecology and Medicine (forthcoming, Academic Press).

5. Lee, E. B. and Markus, L., "Optimal Control for Non-linear Processes", Arch-Rat.Mech.Anal., 8, 36-58 (1968

6. Rink, R. E. and Mohler, R. R., "Completely Controllable Bilinear Systems", SIAM J. Control, 6, 477-486 (1968).

7. Mohler, R. R. and Moon, S. F., "Computation of Optimal Trajectories for a Class of Nonlinear Control Processes", Proc. 1970 IFAC Kyoto Symposium.

8. Mohler, R. R. and Rink, R. E., "Control with a Multiplicative Mode", J. Basic Engineering, 201-206 (1969).

9. Mohler, R. R. and Rink, R. E., "Multivariable Bilinear System Control", Proc. 1968 IFAC Dusseldorf Symposium.

10. Mohler, R. R., "A Common Model in Socio-economics, Ecology and Physiology", Proc. 1971 Int'l Conf. Sys., Man & Cyb., Anaheim, California

11. Hill, T. L. and Kedem, O."Studies in Irreversible Thermodynamics, III. Models for Steady State

and Active Transport Across Membranes", J. Theor. Biol., 10, 399-441 (1966).

12. Snyder, W. S., et al, "Urinary Excretion of Tritium following Exposure of Man to HTO-a two exponential model", Phys. Med. Biol., 13, 547-559 (1968).

13. Langer, G. A. and Brady, A. J., "The Effects of Temperature Upon Contraction and Ionic Exchange in Rabbit Ventricular Myocardium", J. Gen. Physiol., 52, 682-713 (1968).

14. Grodins, F. S., Control Theory and Biological Systems, Columbia University Press, New York, 1963.

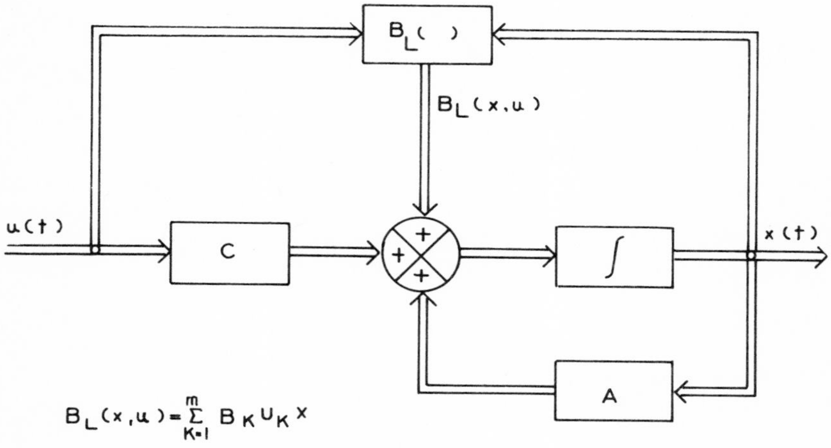

Fig. 1 Bilinear State Diagram

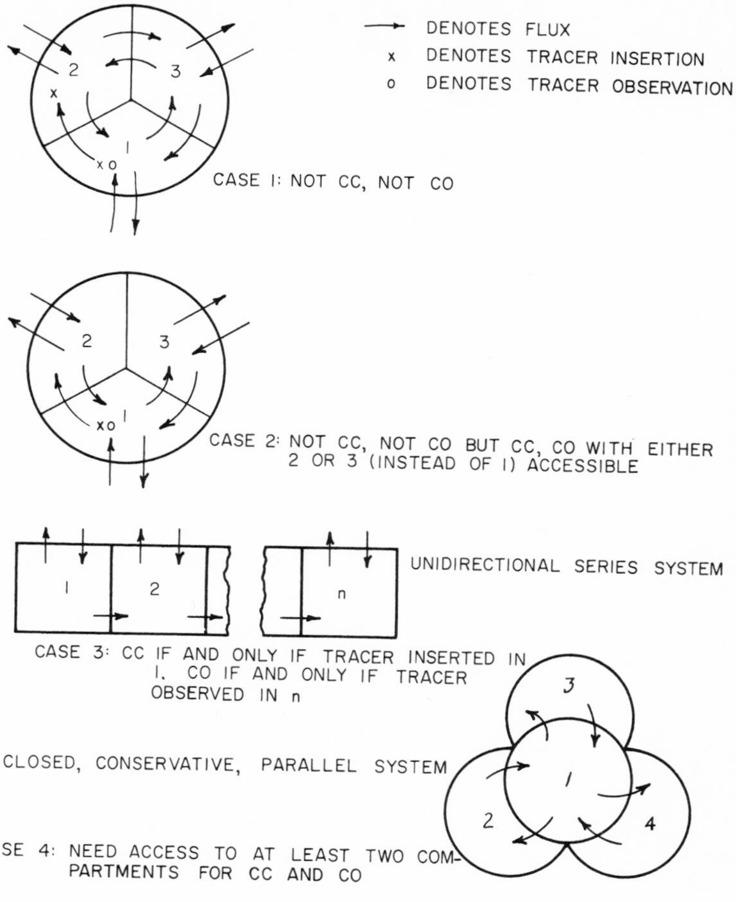
→ DENOTES FLUX
x DENOTES TRACER INSERTION
o DENOTES TRACER OBSERVATION
2
3
x
xo
1
CASE 1: NOT CC, NOT CO
2
3
xo
1
CASE 2: NOT CC, NOT CO BUT CC, CO WITH EITHER 2 OR 3 (INSTEAD OF 1) ACCESSIBLE
1
2
n
UNIDIRECTIONAL SERIES SYSTEM
CASE 3: CC IF AND ONLY IF TRACER INSERTED IN 1, CO IF AND ONLY IF TRACER OBSERVED IN n
CLOSED, CONSERVATIVE, PARALLEL SYSTEM
3
1
2
4
CASE 4: NEED ACCESS TO AT LEAST TWO COMPARTMENTS FOR CC AND CO

Fig. 2 Tracer Accessibility Tests

MATHEMATICAL MODELS AND PROSTHESES FOR SENSE ORGANS (°)

Emanuele Biondi and Roberto Schmid

Istituto di Elettrotecnica ed Elettronica
Centro di Studio per la Teoria dei Sistemi del CNR
Politecnico di Milano
Milan, Italy

Abstract

The general structure of a sense organ is first illustrated, and the principal components are defined. For each component the type of mathematical models which are presently available or can become available in the near future is pointed out by making reference to a particular classification of models of biological systems. The main psycophysical processes which are responsible of some fundamental aspects of the sense organ behaviour are considered and the possibility of their description by means of mathematical models is discussed. Finally, the problem of designing direct and substitutive prostheses for impaired sense organs is approached on the basis of the notions previously introduced and by assuming the possibility of constructing satisfactory mathematical models.

1. Introduction

The advantage of approaching the study of prostheses for impaired sense organs on the basis of mathematical models has been outlined by many authors.

(°) Work supported by C.N.R., Rome, Italy

Models can describe real systems with different degrees of approximation. Even at the level of qualitative block diagrams a model can be useful to suggest solutions other than the obvious one of amplifying the input signals in order to take advantage of the residual functionality of the damaged organ . If the blocks are then specified at a functional level, it will be possible to define the prosthesis quantitatively and to open the way to its future construction.

When unconventional prostheses are employed, subjects need a training period to discover the new code of messages and to become familiar with it. A model for pattern recognition and learning processes in man can suggest to the trainers how to behave in order to make the prosthesis accepted and to obtain the best results.

The aim of this paper is to develop some considerations which can be particularly useful when the problem of designing new prostheses for impaired sense organs is approached in the sense previously outlined. Only those functions which can be recognized in all the sense organs are considered and discussed in their essential aspects. Moreover, in accordance with the purpose of this paper, the coupling between sense and motor organs is not examined in detail, and sensations are the only outputs considered at the cortical level.

With respect to the general topic of the present seminar, particular emphasis will be given to those aspects of the sense organ behaviour leading to models of dynamical systems with variable structure and parameters and to bilinear systems.

Models of biological systems are first classified on the basis of their equivalence to the real systems. Then, the general structure of a sense organ is illustrated, and the principal components are defined. For each component the class of models which are presently available in the literature or can be expected to be available in the near future is pointed out. A deterministic model of neuron is also introduced and later used to explain processes like attention, habituation, fatigue and plasticity in terms of neural nets with variable structures and parameters. Processes for which the experimental evidence mainly consists of psychophysical data are then examined. A special emphasis is given to the processes of pattern recognition and learning. Finally, the problem of designing direct and sub - stitutive prostheses for impaired sense organs is discussed on the basis of the notions developed in the previous sections.

Going through the different sections of the paper, the reader will certainly note some formal discontinuity. This is mainly due to the nature of the problems there treated. In some cases the status of the present knowledge permits a suffi - cient mathematical formalization, in other cases only qualitative considerations can be done.

The reader will also note an unconventional presentation of the bibliography. Since a very large spectrum of problems has been considerd in the paper and most of the problems are only treated in their fundamental aspects, we thought that it was not convenient to make reference to single specific papers. Thus, a bibliography has been prepared where only several books are suggested. Each book is referred to a particular topic considered in the paper, and the reader will find there general notions as well as references to specific papers where single problems are treated in detail.

2. Different types of models of biological systems

It is a common practice to classify models of physical systems according to the nature of the mathematical relationships which form the model. Thus, models belonging to the same class can be examined with the same mathematical techniques.

In the case of biological systems it turns out to be useful to classify models according to their degree of equivalence to the systems they intend to describe. Actually, models can be set up to different goals: to obtain a better knowledge of physiological systems, to express experimental data concisely, to predict the behaviour in situations which are unnatural or unreachable by experiments, to make diagnoses, to find therapeutic schemes or to design prosthetic devices, artificial organs or specific biomedical instrumentations. Depending on the purpose, there is need of different insight into the processes involved, and different approximations to the real situation are to be considered satisfactory.

From this point of view, we suggest to divide models of biological systems in the following five classes.

- anatomo-functional models
- functional models
- abstract-functional models
- abstract models
- artificial models.

- Anatomo-functional models are such to present a complete correspondence between the model structure and parameters and the anatomo-physiology of the modelled biological system. They follow from identification processes "from the inside", which need the definition of the system state and the construction of the input-state and state-output transformations. All parameters appearing in these mathematical relationships should be specified in terms of anatomic or chemico-physical properties of the biological system.

- Functional models are such to present a complete correspondence between model and system external behaviour. They follow from "black box" identification processes, which lead to the construction of input-output relationships. Therefore, a well defined correspondence between the structure and the parameters of the model and those of the biological system does not necessarily exist.

- Abstract-functional models are such to present a complete correspondence with the physical system only for what concerns some aspects of its external behaviour. They are constructed from data obtained under experimental conditions which donot permit a complete identification of the system. To supply the gap in experimental knowledge more or less abstract conjectures are introduced to predict the system behaviour in situations which are unreachable by experiments.

- Abstract models are entirely developed from assumptions based on very poor experimental evidence. They are the result of mathematical constructions from initial axioms more than the formal expression of the experimental analysis.

- Artificial models are mathematical descriptions of artificial devices which perform tasks similar to those carried on by a specific biological organ. The way these tasks are performed may be completely different in the artificial and natural system. Thus, it is in a rather general sense that artificial models can be considered as models of biological systems. Anyway, they can be used as starting point for programming experiments when no anatomic nor physiological background exists.

The classification so far introduced is somewhat arbitrary since no precise distinction can be made between models belonging to contiguous classes.

The possibility of obtaining a model of a given type depends on several factors: the complexity of the system considered, the kind of experimental data which are available and the nature of the experiments which can be used to confirm the model. Preliminary models may be abstract models. Then, a reasonable correspondence can be established between the functions included in the model and those performed by spatially defined regions of the biological system. As long as the theoretical and experimental analysis proceeds, it may be possible to identify most of the variables and parameters with biophysical or biochemical entities and to obtain anatomo-functional models.

3. Structure and components of sense organs.

Sense organs perform a twofold function. First of all, they transform external stimuli into sensations perceived at cortical level. Secondly, they produce appropriate inputs to other systems, e.g. to the postural control system.

The general structure of a sense organ is presented in Fig. 1.

The peripheral transducers convert external input signals into appropriate stimuli to sensory cells. Obviously, their characteristics depend on the nature of the signals to be converted. Each sense organ has a specific type of peripheral transducer.

Without going into details, which would lead us to consider the sense organs separately, it is possible to say that the experimental conditions at the level of the peripheral transducers are in general such to permit the construction of anatomo-functional models.

The sensory receptors transform the output of the peripheral transducers into electrical signals. In many cases the mechanism of transformation is far to be clear, while sufficient data are available for what concerns the static and dynamic relationships between input and output. Thus, only functional models can be constructed.

It is usually said that sensory receptors show adaptation. If the word adaptation is used to mean a declining response to a constant stimulation, its use appears to be improper. Actually, it is possible to account for this fact by assuming the presence of zeros into the transfer function, which describes the dynamics of the linear part of the system. There is no need of adaptive nor time variant models.

As a matter of fact, a sort of adaptation exists in many sensory receptors their characteristics being adapted to the input. In the case of the auditory system, the polarization of the cochlear sensory cells seems to be controlled through the efferent pathways and adapted to the mean value of their input, (outer feedback in Fig. 2). At the same time, another control loop would adapt the differential gain of the coclear receptors to the amplitude of sound (inner feedback in Fig. 2). This could explain the tremendous range of linearity of the overall auditory system with respect to the small range of linearity of the sensory cells (1). Furthermore, the system would work always with maximal sensitivity without risk of damages.

It is worth noting that such an adaptation of the receptor gain may lead in general to the construction of special kind of bilinear models.

Assume in the diagram of Fig. 2 $z_1 = u_o$. Moreover, let the dynamics of the sensory receptors be described by the first order linear differential equation

$$\Delta\dot{x}(t) = -a\,\Delta x(t) + \Delta w(t) \tag{1}$$

and let

$$\Delta y(t) = c\,\Delta x(t) \tag{2}$$

with a and c positive constants. Denoting the slope of the linear part of the static characteristic by $\mu(\Delta y)$, we can write

$$\Delta w(t) = \mu(\Delta y)\,\Delta u(t). \tag{3}$$

Obviously, eq. (3) holds as long as $|\Delta u| < u_s$ where u_s is the value of the input signal for which the static characteristic saturates. Due to the presence of the external loop it can be reasonably assumed that the condition on $|\Delta u|$ is always satisfied.

If μ_o is the slope corresponding to receptor maximal sensitivity, we can also assume

$$\mu(\Delta y) = \mu_o - \frac{\mu_1}{c}\,|\Delta y| \tag{4}$$

with μ_o and μ_1 positive constants. Then, from (1), (2), (3) and (4) it follows:

$$\Delta\dot{x}(t) = -a\Delta x(t)+\mu_o\Delta u(t)-\mu_1\Delta u(t)\;|\Delta x(t)| \tag{5}$$

$$\Delta y(t) = c\,\Delta x(t)$$

(1) Obviously, the adaptation of the system differential gain introduces a nonlinearity. However, this nonlinearity can be easily compensated at the level of the C.N.S. where the signal $z_2(t)$ is of course available.

Let $\Delta_1 X$ and $\Delta_2 X$ be the sets of non negative and negative real numbers respectively.

From (5) it follows

$$\Delta\dot{x}(t) = -a\Delta x(t)+\mu_o \Delta u(t)-\mu_1 \Delta u(t)\Delta x(t)$$

$$\Delta y(t) = c\ \Delta x(t)$$

for $|\Delta u| < u_1$ and $\Delta x \in \Delta_1 X$, and

$$\Delta\dot{x}(t) = -a\Delta x(t)+\mu_o \Delta u(t)+\mu_1 \Delta u(t)\Delta x(t)$$

$$\Delta y(t) = c\ \Delta x(t)$$

for $|\Delta u| < u_s$ and $\Delta\ x \in \Delta_2 X$.

In each subregion $\Delta_1 X$ and $\Delta_2 X$ the system is described by a bilinear model. Eqs. (5) can be considered as a two-mode bilinear model of adaptive sensory receptors.

The nuclei are neural nets, specialized for some function. We can distinguish between two classes of neural collections: intermediate nuclei and cortical sensory areas. The intermediate nuclei produce inputs to other nuclei, while cortical sensory areas produce both inputs to other nuclei and sensations.

To the purpose of this paper it turns out to be useful to divide intermediate nuclei into specific nuclei and interaction nuclei. Specific nuclei belong to a particular sense organ, and they have no direct signal exchange with other systems. Interaction nuclei also belong to a particular sense organ, but they interact with other systems [2].

Fig. 3 shows a possible organization of specific and interaction nuclei.

Although neuroanatomy and neurophysiology have made a tremendous progress in the last few years, it is not yet possible to establish the detailed organization of neurons inside a nucleus for some certainty. Thus, no sufficient knowledge exists

(2) This classification may appear incorrect from a strict neurophysiological point of view. It has been introduced to make easier the following discussion on prostheses.

for constructing anatomo-functional models.

Also the construction of functional models seems to be extremely hard. In many cases, it is not clear which is the code of the information contained in the input and output signals of a nucleus, and it is almost impossible to determine how information is treated by the nucleus itself. The existence of many feedbacks from higher nervous centers makes the analysis more complex. On the other hand, chemical and surgical suppression of these feedbacks can deteriorate the nucleus and reduce the significance of the results so obtained.

The status of the present knowledge and the objective limits of the experimental research in neurophysiology are such that one can reasonably expect to construct only abstract or abstract functional models of nucli. A methodological scheme may be the following one: to set up a functional model of the neuron on the basis of neurophysiological considerations, then to make reasonable assumptions on the structure of neural interconnections in a nucleus, and finally to use the mathematico-deductive method to establish the functions that a nucleus can carry out. Further experiments, suggested in part by the previous theoretical analysis itself, may reveal the inadequacy of the original model of the neuron and suggest to modify the assumptions on neural interactions.

An alternative procedure for obtaining a mathematical description of the neural processes in sense organs will be outlined in section 6, where models based essentially on psychophysical data are discussed.

It is beyond the purpose of this paper to review the large literature on models of neurons and neural nets. We shall only give a brief description of a deterministic model of neuron, which represents the first step of an identification process from the inside and makes reference to a precise definition of dynamical system. This model will be later used to introduce some comments on processes like attention, habituation, fatigue and plasticity.

4. A deterministic model of neuron

Without going into a neurophysiological description of neuron behaviour, we can summarize some conclusions leading to a model which takes into account only the main characteristics of certain classes of neurons.

A neuron is a dynamical system defined by the following sets and transformations.

1) The set of time $T = \{t\}$. Leaving the assumption of neuron complete synchronization, we can define T by letting

$$T = |0,+\infty) = \text{the set of non negative real numbers.}$$

2) The set of inputs $U = \{\underline{u}\}$. Let m be the number of dendrites of the neuron.
Since the impulses reaching the synapses of the dendrites can be conveniently treated as unitary impulses with different time distribution, the input vector $\underline{u}$ can be represented by an ordered m-ple of 0 and 1's. Then, U is a finite set of 2^m elements. Each element corresponds to a vertex of the m-cube, that is

$$U = \{\text{vertices of the m-cube}\}$$

3) The set of input functions $\Omega = \{\underline{u}(\cdot)\}$. Since any sequence of impulses can reach the synapse of a dendrite (3), we assume that the vectors $\underline{u}(\cdot)$ are ordered m-ples of binary (0,1) functions $u_i(\cdot)$ zero almost everywhere. This defines the set Ω.

4) The set of states $X = \{\underline{x}\}$. We can distinguish among three different working conditions of neurons: excitation, recovery and refractoriness. During excitation the neural potential ξ varies with time. If ξ exceeds a threshold Θ at time $\bar{t}$, the neuron fires an impulse along its axon. The neural potential ξ drops to zero (recovery) and it remains equal to zero for a period T_{rf} corresponding to the neuron

(3) Actually, there are some restrictions on the minimal interval between two impulses(see points 5 and 6 of this section).

refractory period. The knowledge of ξ is not sufficient to define the state of the neuron. During the refractory period we must specify the time τ leaving to the end of this period. Then, the neuron can be described as a second order dynamical system with state $\underline{x}$ defined by the pairs(ξ,τ). The set X is given by

$$X = \{(\xi,\tau):(\xi,\tau) \in X_1 \cup X_2 \cup X_3\}$$

where the subregions X_1, X_2, and X_3 are specified in Fig. 4 [4].

5) The set of outputs $Y = \{y\}$. The output of the neuron is the sequence of impulses fired along its axon. The minimal interval between two impulses is $T_{rc} + T_{rf}$. The set of outputs is the finite set

$$Y = \{0,1\}.$$

6) The set of output functions $\Gamma = \{y(\cdot)\}$. The set Γ is the set of the binary (0,1) functions zero almost everywhere and satisfying the condition that $y(\cdot)$ cannot be equal to 1 at two instants apart less than $T_{rc} + T_{rf}$.

7) The input-state transformation $\phi : X \times U \to X$. The input state transformation is defined by the following state equations. Excitation:

$$\dot{\xi}(t) = -a\ \xi(t) + \sum_{1}^{m}{}_i\ w_i u_i(t)\ \delta(t)$$

$$\dot{\tau}(t) = 0 \qquad\qquad (\xi,\tau) \in X_1 \quad (6)$$

where a is a positive constant related to the period of latent summation of the neuron, w_i are the weights associated to the input synapses ($w_i > 0$ if the synapse is excitatory, $w_i < 0$ if the synapse is inhibitory), and $\delta(t)$ if the Dirac function.

Recovery:

$$\dot{\xi}(t) = -\frac{\theta}{T_{rc}} \qquad \dot{\tau}(t) = \frac{T_{rf}}{T_{rc}} \quad (\xi,\tau)\in X_2 \qquad (7)$$

---

(4) For the sake of simplicity we have assumed that during the recovery period ξ drops to zero linearly in a time T_{rc}.

Refractoriness:

$$\dot{\xi}(t) = 0 \qquad (\xi,\tau) \in X_3 \qquad (8)$$
$$\dot{\tau}(t) = -1$$

It should be noted that the final state reached at the end of a working condition represents the initial state for the next working condition. There is no discontinuity in the values of the state along a trajectory.

7) <u>The state-output transformation</u> $\eta: X \to Y$. The state-output transformation is given by

$$y(t) = \delta_{\theta}^{\xi(t)} = \begin{cases} 0 & \text{for } \xi(t) \neq \theta \\ 1 & \text{for } \xi(t) = \theta \end{cases}$$

As long as the m+4 parameters $w_1,\ldots,w_m,a,\theta$, T_{rc} and T_{rf} are assumed to be constant, this model classifies the neuron as a second order, time invariant, nonlinear dynamical system. The set X, which is not a vector space, can be devided into three disjoint subregions. Inside each subregion the movements of the state are solutions of different linear differential equations. The system has only one equilibrium state, the null state (0,0), which turns out to be stable. Finally, the system is not reversible, fully state controllable and reachable, but not differentially.

The assumption of neuron constant parameters cannot be accepted for many reasons. To the purpose of this paper, fluctuations following changes of internal environmental conditions (body temperature, methabolism and so on) can be conveniently neglected. On the contrary, this assumption must be removed if a model of neural nets accounting for processes like attention, habituation, fatigue and plasticity is to be set up. These processes are briefly discussed in the next section.

5. Attention, habituation, fatigue and plasticity.

Attention is the process through which nucleus sensitivity to incoming signals is increased by reducing the inhibitory feedbacks from the cortical sensory areas. It can be observed by measuring nucleus activity in alert and non alert animals.

Habituation is the process through which nucleus activity in response to the same stimulus decreases when the stimulus is repeated many times.

Fatigue is the process through which nucleus activity decreases when a time varying stimulus lasts for long.

All these processes can be easily explained by assuming the possiblity of changing the neuron threshold θ or the weights w_i or both. Attention may be the result of a threshold decrease, while habituation and fatigue the effect of a threshold increase.

Since the dynamics of these processes is much slower than neuron dynamics, it is almost impossible to establish from input-output considerations if the decrease (increase) of the output discharge in response to the same input is due to an increase (decrease) of the threshold or to an equal decrease (increase) of all the weights w_i. At steady state the effect on the system output is exactly the same.

Only neurophysiological considerations may help to solve this problem. From a mathematical point of view, it is worth noting that the linearity of the state equations (6) is not lost, if only the threshold θ is controlled, as shown by the block diagram of Fig. 5 which is referred to the excitatory period. On the contrary, eqs. (6) become nonlinear if the weights w_i are controlled.

In the general case, neurons can be described as time invariant dynamical systems with the input vector $|u_1 \ldots u_m \; w_1 \ldots . w_m, \; \theta|$ as shown in Fig. 6, which corresponds to the excitatory period.

To the aim of this paper it is worth mentioning another process which can be described only by models of neural nets with variable structure and parameters. This process, referred to as plasticity, consists of a reorganization of a neural net such that some perceptive properties of a sense organ, which have been lost for a damage in some part of the nervous system, are partially regained.

The processes discussed in this section should

be described more properly by means of multilevel hierarchical models in which each level performs an adaptive control of the structure and parameters of the lower level.

Since all modifications of the structure and parameters of the neural nets produce an effect at psychophysical level, the models of neural nets proposed to explain attention, habituation, fatigue and plasticity can be tested by using the results of specifically oriented psychophysical experiments. Thus, the methodological scheme suggested at the end of section 3 can be completed as shown in Fig. 7.

6. Psychophysical processes

In the previous section some processes which can be investigated by both neurophysiological and psychophysical experiments have been discussed. However, the overall performance of a sense organ strongly depends on central processes which, at present, can be analysed almost only through psychophysical tests.

The most relevant among these processes are listed below:

- Discrimination (distinction between two external stimuli).
- Relative scaling (ordering of external stimuli according to their relative magnitude).
- Absolute scaling (assignment of a value to the magnitude of an external stimulus).
- Classification (distinction of different classes of patterns of external stimuli).
- Pattern recognition (assignment of patterns of external stimuli to their own classes).
- Learning (taking advantage from previous experiences to carry on some process in a more efficient way).

With the exclusion of learning, which is essentially a dynamical process all the other processes can be considered at steady state or during transients. Consequently, two classes of models can be constructed: models for stationary processes and models for dynamical processes.

By using communication and information theory as theoretical support it is fairly easy to set up artificial models for discrimination, absolute and relative scaling, classification and pattern recognition considered as stationary processes. The results of psychophysical tests are in general verv simple as long as strongly state conditions are examined, and they can be fitted by artificial models quite satisfactorily.

More realistic models have been also tempted. They take into account the static characteristics of the peripheral transducers and sensory receptors and the static properties of neural networks with threshold elements like the neuron described in section 4. With few exceptions, these abstract-functional models fit the results of the psychophysical tests in a qualitative more than quantitative way.

For what concerns the dynamical aspects of these processes, they all point out the necessity of describing the neural systems which perform these processes by means of models with variable structure and parameters.

In the next section the processes of pattern recognition and learning are discussed jointly, and the principal elements of a sense organ involved in these processes are briefly illustrated.

7. Pattern recognition and learning

Learning is an adaptive process which includes pattern recognition as one of its major goals. A process has been defined by Tsypkin to be adaptive when the parameters and the structure of a system are adjusted so as to use accumulation of incoming information in such a way to achieve some specified goal.

The learning problem in pattern recognition can be stated as that of using a finite number of observations of representative members of different classes of pattern to construct a surface separating a multidimensional space into domains corresponding to the classes. Once the learning is completed, recognition reduces to an investigation of the domain to which an observed pattern belongs.

The processes of pattern recognition and learning in biological systems have been the object of many investigations since the time Wiener was developing the basic ideas of Cybernetics and probably far before although in a less formal way. Many theories have been proposed, while the difference between Cybernetics and Artificial Intelligence Science was becoming clearer and clearer. On one side, there are people attempting to explain how human brain can perform particular tasks. On the other side, there are people attempting to determine how the same tasks can be performed efficiently by machines, without the purpose of deliberately imitating the human behaviour and refusing the principle that brain-like solutions are the optimal ones also for computing machines.

According to the definition given in section 2, the solutions proposed by the artificers of pattern-classifying and learning machines can be considered as artificial models for the natural processes of pattern recognition and learning. Once the misunderstanding of presenting good designs of computing machines as biological models of the human brain has been removed, artificial models can be used in the way previously outlined, that is to program specific psychophysical or neurophysiological experiments, and to make rational hypotheses in the case of lack of experimental evidence. If the contributions from anatomy physiology, psychophysics and artificial intelligence science are considered altogether, the problem of constructing a mathematical model of a sense organ can be illustrated by the diagram of Fig. 8.

It is not our purpose to review the models of the human brain set up in the last years nor to illustrate the many artificial models for the processes of pattern recognition and learning. We shall only make reference to the previous statement of the learning problem in pattern recognition and to Tsypkin's definition of adaptive system in order to identify the principal elements of a sense organ involved in these processes.

Two cascade elements are first of all necessary to recognize and classify external stimuli : a feature extractor and a pattern recognition system.

The feature extractor must be a neural network which decodes the messages sent by the peripheral nervous system and computes those parameters which will permit the identification of the external stimuli (it measures single fiber discharge, counts the number of active fibers, correlates fiber activity to fiber position, computes mean values, carries out frequency analysis, and so on). The feature extractor must be defined by its static and dynamic properties. The number of parameters computed by the feature extractor neural network determines the dimension of the space where pattern recognition is then carried on.

The pattern recognition system must be a neural network which recognizes elements in the output space of the feature extractor. It is essentially a static system, which uses neurons as threshold elements to contruct the surface separating the output space of the feature extractor into domains corresponding to the various classes of pattern and to find elements in these domains.

In order to solve the learning problem in pattern recognition three basic elements are necessary: a long term memory, a short term memory and a decision maker.

The long term memory must be a neural network where the main characteristics of the feature extractor and pattern recognition system are stored.

The short term memory must be a neural network where the input messages and the outputs of the feature extractor and the pattern recognition system are stored for a short period of time.

The decision maker examines the content of the short and long term memory and the answer of the external world. It decides how to modify the characteristics of the feature extractor and the pattern recognition system in order to improve the overall system performance.

The behaviour of the feature extractor and that of the pattern recognition system can be modified by changing the structure of neuron interconnections or/and the values of neuron parameters. It is a widespread opinion that a certain degree of randomness initially exists in neuron interconnections, which is reduced as long as the learning process proceeds. On the other hand, it has been proved that the property of pattern recognition of networks of elements behaving like neurons (perceptrons) can be improved by simple rules of reinforcement of the weights associated to the inputs.

Fig. 9 shows schematically the organization of the elements necessary to carry on the processes of pattern recognition and learning.

8. Prostheses for impaired sense organs

Let us consider the diagram of Fig.10, which is the result of the analysis carried on in the previous sections, and assume that the sense organ A is damaged, while the sense organ B behaves normally. Two kinds of prostheses can be devised:

a) Direct prostheses, which make use of the input channel of the sense organ A to send information about signals specific to the same sense organ.
b) Substitutive prostheses, which make use of the input channel of the sense organ B to send information about signals specific to the sense organ A (e.g. the tactile system is used to send visual or acoustical messages).

In the case of direct prostheses, the optimal design problem is reduced to that of determining how to process the natural input signals in order to maximize the amount of information contained in the output of the prosthesis-impaired sense organ system.

If we consider the scheme in Fig. 11, we can define at least two errors, which can characterize the difference between the behaviour of the normal and impaired sense organ. For the sake of simplicity it is assumed that the lesion is located at the level of the transmission system.

The error E_1 can be in general determined quite easily. In fact, its determination requires the knowledge of models of the peripheral part of the system, for which sufficient experimental data are generally available. Unfortunately, the error E_1 is not the most significant one. Prostheses designed to minimize E_1 can introduce a coding of the external messages such that the central nervous system is then completely unable to carry on the task of decoding and interpreting the messages. The failure of some acoustical prostheses based on the minimization of E_1 may be explained in this way.

The error E_2 is undoubtely the most important one, since it defines the difference between the pattern recognition made by the normal and the impaired subject. At present, no satisfactory models are available for estimating E_2. However, on the basis of the present knowledge the following trial and error process can be started.

An uncomplete mathematical model of the sense organ is first constructed, and a prosthesis can be designed in order to minimize E_2 according to the model. As long as new pieces of information about the system behaviour are obtained, the characteristics of the model and those of the prosthetic device are modified accordingly. The experiments which can be planned by the use of the prosthesis may be of great help to obtain a better knowledge of the system.

Also the experiments made by using direct prostheses on normal subjects may be extremely important. As a matter of fact, any manipolation of the natural messages introduced to facilitate understanding by an impaired subject, makes understanding more difficult for a normal subject. On the other hand, experiments tend to suggest that, in general, if a normal subject is unable to understand the new coded messages after a brief period of training, the impaired subject will fail, too.

The problem of designing substitutive prostheses can be stated in almost the same way. Obviously, both the impaired and the substitutive sense organ should be described by means of mathematical models, although different approximations may be necessary. In particular, if the two organs interact each other, the behaviour of the interaction nuclei must be examined in detail. Useful suggestions can be derived on the way to process input signals specific to the impaired sense organ before they are sent to the substitutive organ. If the lesion is set up in the peripheral part of the system, it can be bypassed by sending throughout the interaction nuclei meaningful messages to the pattern recognition and learning systems of the peripherally damaged sense organ.

If the interaction between the impaired and the substitutive sense organ is so weak that the first one is practically excluded from both the transmission and the interpretation of messages, or in the case of lesions beyond the interaction nuclei, a detailed model is needed only for the substitutive organ.

Attempting to construct this model, one has to take into account the following fact.

A sense organ may behave quite differently when it receives and processes patterns of signals which are natural for it, and when it has to deal with patterns specific to other sensory systems. Anyway, substitutive sense organs of impaired subjects show an high degree of nervous system adaptation either for parameters or for structure (plasticity). As a metter of fact, the experience proves that an impaired subject has the opportunity of modifying the substitutive sense organ in such a way to make it suitable to the transmission and the processing of messages concerning the impaired organ. In other words, the substitutive sense organ of an impaired subject can modify itself in order to admit as meaningful input signals even those patterns that are for a normal subject significant only for their specific organ.

In consequence of this particular behaviour,

tests of substitutive prostheses performed on normal subjects are completely meaningless.

9. Conclusion

An approach to the study of unconventional prostheses for impaired sense organs, based on the use of mathematical models, has been outlined.

As far as direct prostheses are concerned, this approach can be summarized in the following steps.

1) Set up a mathematical model of the sense organ considered. Even if roughly approximated, the model can suggest new types of direct prostheses.
2) Test the new prosthesis on normal subjects. Prostheses no passing these tests should be rejected, since they will probably fail also with impaired subjects. In this case, turn back to point 1 and try to improve the model by going deeply inside the anatomy and physiology of the sense organ considered. Otherwise,
3) Test the prostheses on impaired subjects. Either positive or negative results can be used to improve the preliminary model of the system. Also the classification of the impaired subjects can be improved by interpreting the results in terms of structure and parameters of the model.
4) If an improvement of the mathematical model is obtained, try to modify the prosthesis accordingly, and repeat the procedure from point 2.

This approach, which could appear somewhat unrealistic, has been followed to design a new direct prosthesis for profoundly deaf subjects. Although good results have been already achieved, this prosthesis will be probably improved as long as new experiments are carried on and the supporting mathematical model is completed.

As far as substitutive prostheses are concerned, a similar procedure may be followed. According to the remarks concluding section 8, tests of substitutive prostheses on normal subjects are meaningless.

Bibliography

Physiology and Psychology.

1. V.B. Mountcastle, Medical Physiology, The C.V. Mosby Co., Saint Louis (1968).
2. W.R. Lowenstein, Principles of Receptor Physiology, Handbook of Sensory Physiology, Vol. 1, Springer Verlag, Berlin (1971).
3. W.A. Rosenblith, Sensory Communication, The M.I.T. Press, Cambridge, Mass. (1961).
4. I.C. Whitfield , The Auditory Pathway, Edward Arnold (Publishers) Ltd., London (1967).
5. R. Plomp and G.F. Smoorenburg, Frequency Analysis and Periodicity Detection in Hearing, A.W. Sijthoff, Leiden (1970).
6. I.M. Gelfand, V.S. Gurfinkel, S.V. Fomin and M.C. Tsetlin, Models of the Structural-Functional Organization of Certain Biological Systems, The M.I.T. Press, Cambridge, Mass. (1971).
7. G. Horn, and R.A. Hinde, Short-term Changes in Neural Activity and Behaviour, Cambridge University Press (1970).

Cybernetics

8. N. Wiener, Cybernetics, J. Wiley, N.Y. (1948)
9. M.C. Yovits, G.T. Jacobi and G.D. Goldstein, Self-Organizing Systems, Spartan Books, Washington (1962).
10. N. Wiener and J.P. Schadé, Nerve,Brain and Memory Models, Progress in Brain Research, Vol. 2, Elsevier Publishing Co., Amsterdam, (1963).
11. N. Wiener and J.P. Schadé, Cybernetics of the Nervous System, Progress in Brain Research, Vol. 17, Elsevier Publishing Co.,Amsterdam (1965).
12. M.A. Arbib, Brains, Machines and Mathematics, McGraw-Hill, N.Y. (1964).

13. E.R. Caianiello, Neural Networks, Springer Verlag, Berlin (1968).

14. C. Cherry, On Human Communication, The M.I.T. Press, Cambridge, Mass. (1957).

15. J.A. Sweets, Signal Detection and Recognition by Human Observers, J. Wiley, N.Y. (1964).

16. J.J. Gibson, The Senses Considered as Perceptual Systems, George Allen and Unwin Ltd., London (1968).

17. D.A. Norman, Models of Human Memory, Academic Press, N.Y. (1970).

Artificial Intelligence Science.

18. N.J. Nelsson, Learning Machines, McGraw-Hill, N.Y. (1965).

19. M. Minsky and S. Papert, Perceptrons, The M.I.T. Press, Cambridge, Mass. (1969).

20. O.J. Grüsser and R. Klinke, Pattern Recognition in Biological and Technical Systems, Springer Verlag, Berlin (1971).

21. N.J. Nelsson, Problem-solving Methods in Artificial Intelligence, McGraw-Hill, N.Y. (1972).

Bioengineering

22. J.L. Flanagan, Speech Analysis, Synthesis and Perception, Academic Press, N.Y. (1965).

23. H.P. Schwan, Biological Engineering, McGraw-Hill, N.Y. (1969).

24. J.H. Milsum, Biological Control Systems Analysis, McGraw-Hill (1966).

25. L. Stark, Neurological Control Systems, Studies in Bioengineering, Plenum Press, N.Y. (1968).

The prosthesis for profoundly deaf people mentioned in section 9 has been first described in: E.Biondi and L.Biondi, Alta Frequenza, 8, 37(1968).

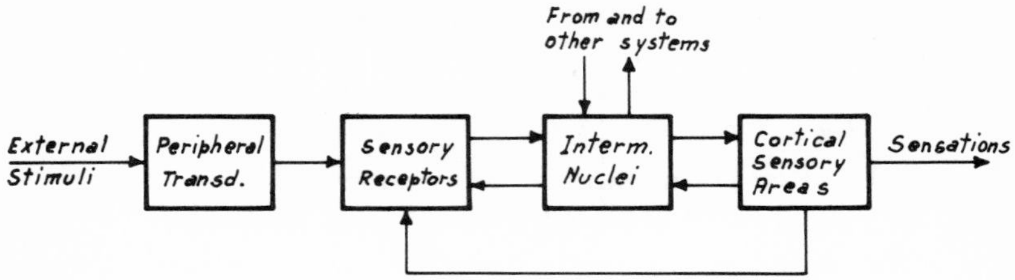

Fig. 1

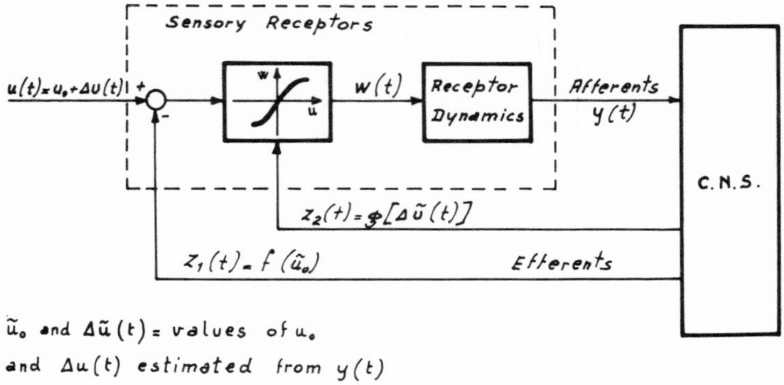

Fig. 2

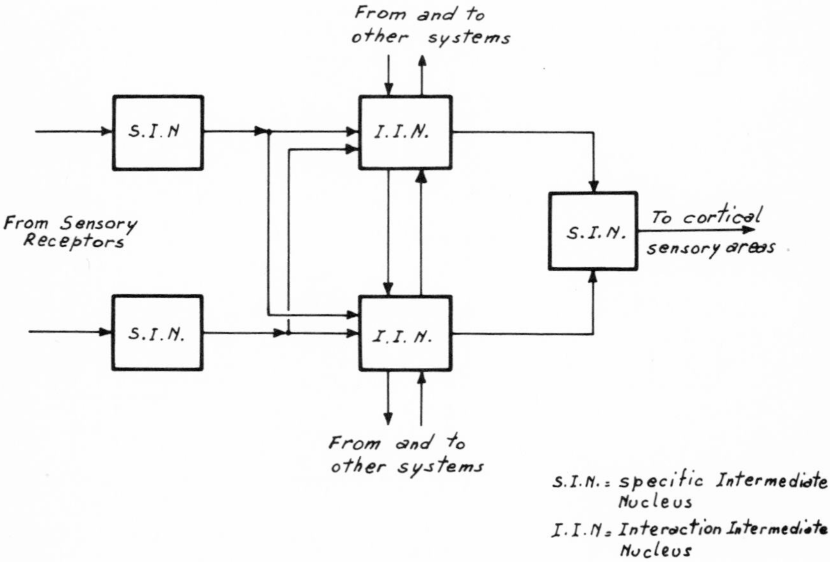

Fig. 3

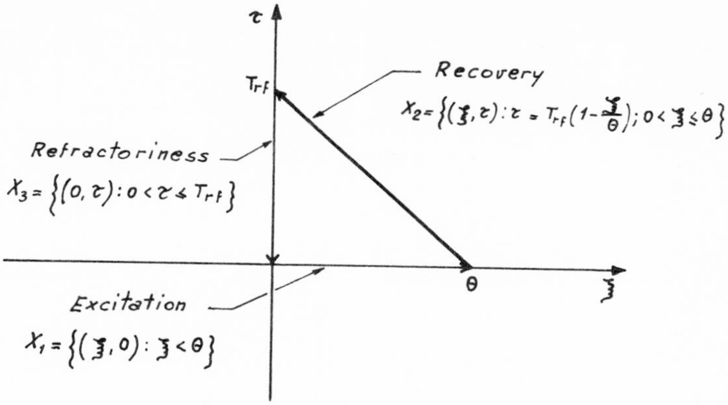

Fig. 4

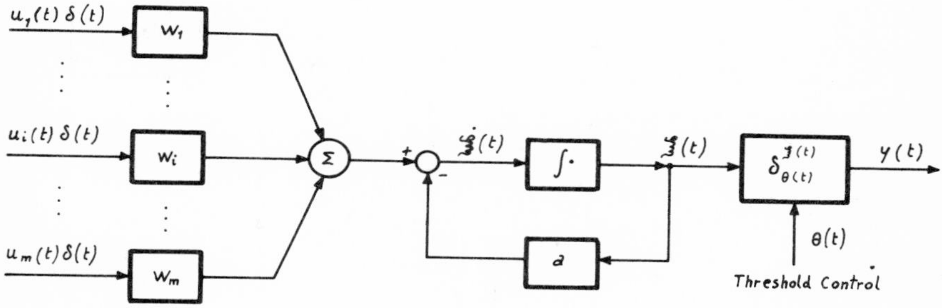

Fig. 5

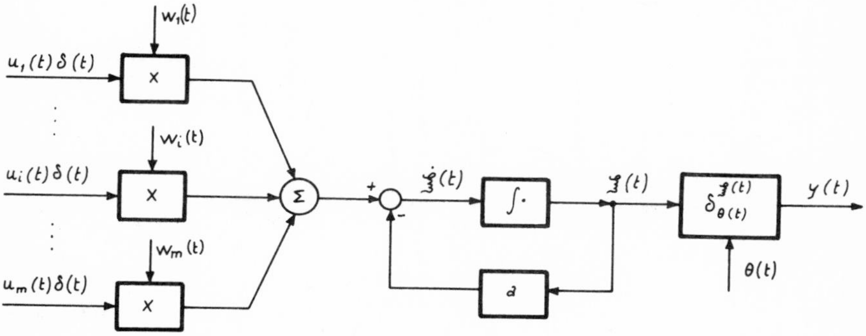

Fig. 6

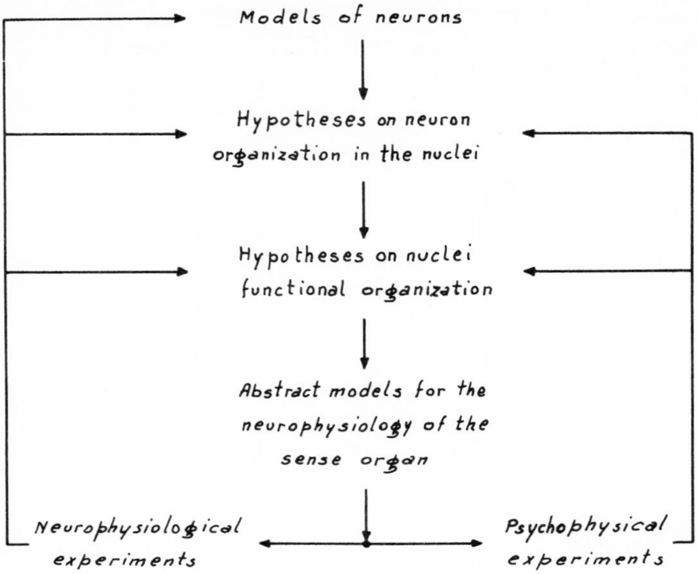
Models of neurons
Hypotheses on neuron organization in the nuclei
Hypotheses on nuclei functional organization
Abstract models for the neurophysiology of the sense organ
Neurophysiological experiments
Psychophysical experiments

Fig. 7

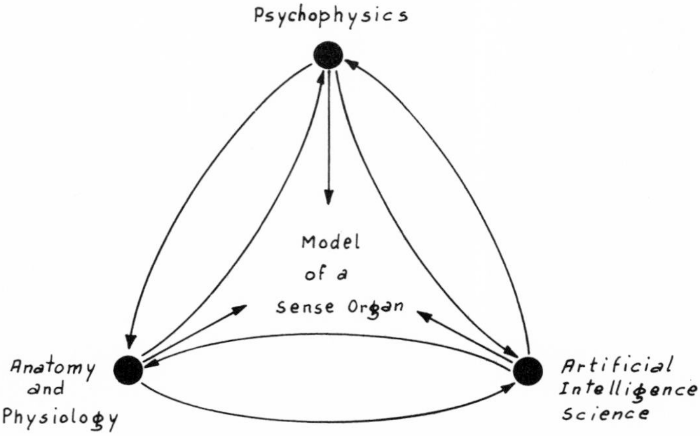
Psychophysics
Model of a Sense Organ
Anatomy and Physiology
Artificial Intelligence Science

Fig. 8

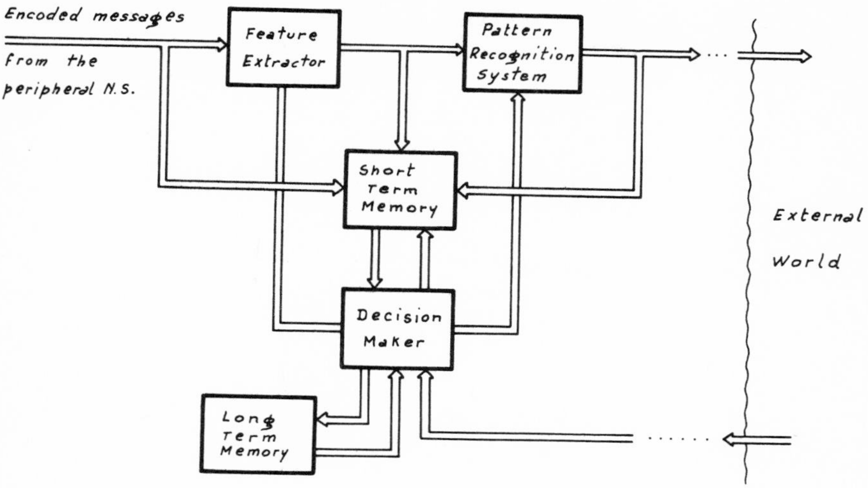
Encoded messages
from the
peripheral N.S.
Feature
Extractor
Pattern
Recognition
System
Short
Term
Memory
Decision
Maker
Long
Term
Memory
External
World

Fig. 9

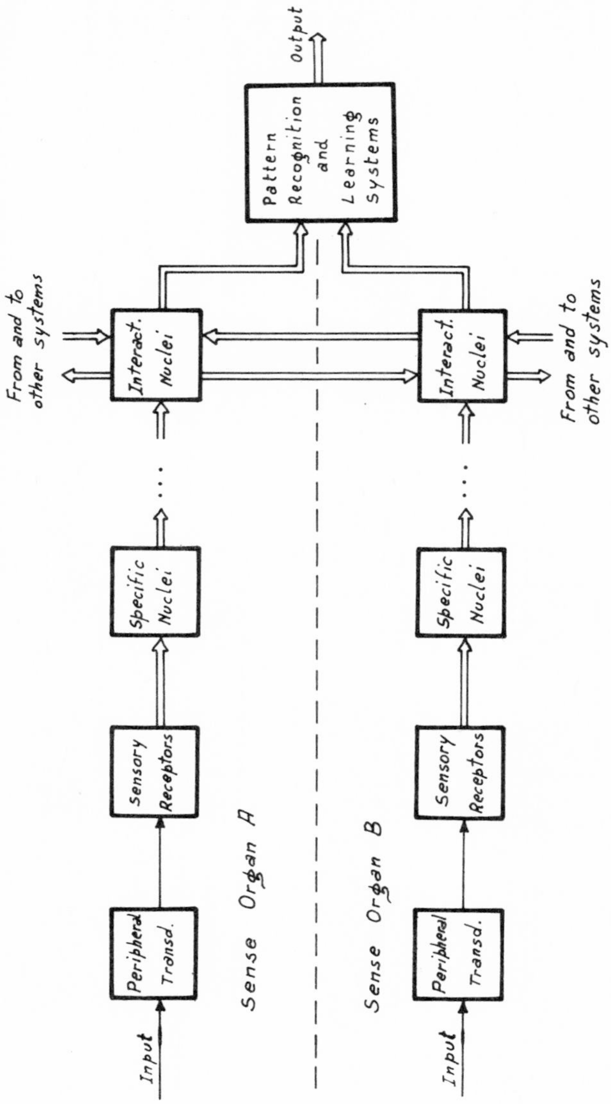
Input
Peripheral Transd.
Sensory Receptors
Specific Nuclei
Interact. Nuclei
From and to other systems
Pattern Recognition and Learning Systems
Output
Sense Organ A
Sense Organ B

Fig. 10

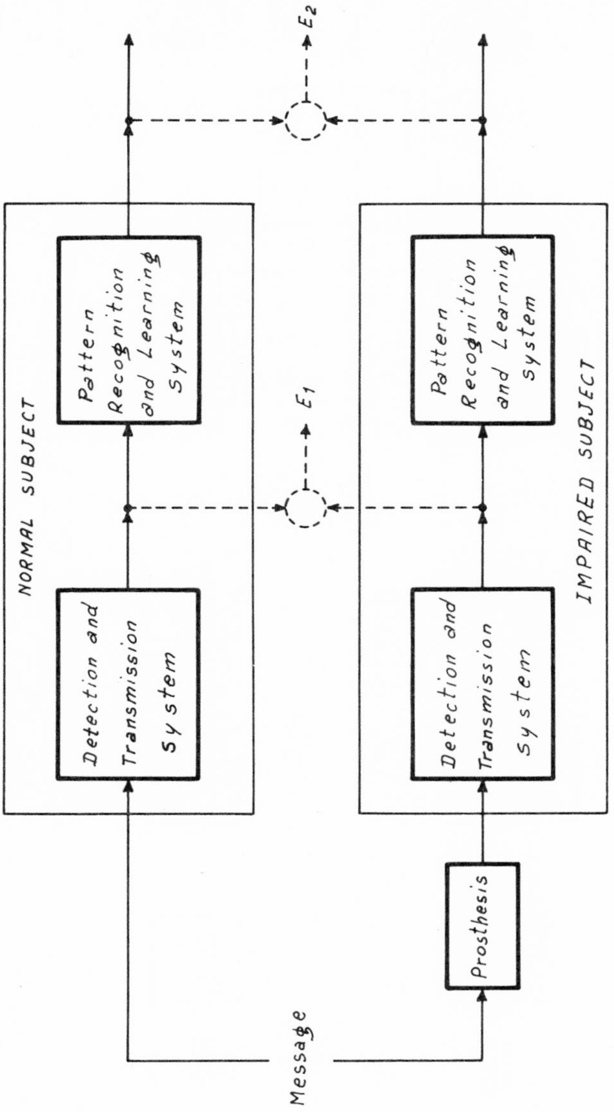
NORMAL SUBJECT
Detection and Transmission System
Pattern Recognition and Learning System
IMPAIRED SUBJECT
Prosthesis
Detection and Transmission System
Pattern Recognition and Learning System
Message
E_1
E_2

Fig. 11

VARIABLE-STRUCTURE ASPECTS OF ECOSYSTEMS

Bernard C. Patten

Department of Zoology and Institute of Ecology
University of Georgia, Athens, Georgia 30601, U.S.A.

I. Introduction

Ecosystems will present an important, if not definitive, challenge to systems analysis and simulation as evolved for physical systems. Goals of ecology may be similar to engineering goals, such as optimal design and control, perhaps even synthesis. But also they may be different, emphasizing scientific discovery, explanation and prediction, and not just manipulation. The scope and complexity of ecosystems is unfamiliar in the experience of physical science, but familiar in biology. Therefore, it seems certain that current concepts and methods of systems science will undergo profound alteration as attention turns to systems of macro scale.

The first challenges will be modeling, particularly where detail is to be preserved. The ultimate challenges, it seems, will be in the convergence of theory and practice, in the development of systems analysis modes that are "model free" in the sense of yielding information more directly related to real-world systems than to their mathematical homomorphs.

This paper is directed to both ends of this spectrum. Its purposes are (1) to review briefly some general problems of modeling ecosystems, and (2) to give an example of a systems analysis concept (sensitivity) modified in the model-free direction to suit ecosystems.

II. Problems in Total Ecosystem Modeling

The enormous complexity of ecosystems demands that

University of Georgia, *Contributions in Systems Ecology*, No. 7.

their modeling and analysis be a joint enterprise between ecologists and systems scientists. The systems scientist must provide a sufficient theory of modeling (e.g., Ziegler, 1970, 1971a, 1971b; Ziegler and Weinberg, 1970) and guidelines for putting it into practice. The theory must provide for decoding back to a real-world frame of reference, the inverse modeling problem (Patten, 1970). The ecologist must contribute a conceptualization that fits the theory based on knowledge that only he possesses of the prototype ecosystem. As an example of what this conceptualization may involve, Table I lists components for a total ecosystem model of a deciduous forest watershed in western North Carolina. The model is to represent processes of mineral cycling in biotic and abiotic sectors of the watershed ecosystem. It took approximately 25 ecologists and graduate students five months to resolve known information about the watershed into the Table I categories. The 202 compartments are somewhat less than 20 percent connected so that there are approximately 8000 pairwise couplings in the system interconnection. The model to be developed will take the form

$$\begin{aligned} \dot{\underset{\sim}{x}}(t) &= A(t)\underset{\sim}{x}(t) + B(t)\underset{\sim}{u}(t), \\ \underset{\sim}{y}(t) &= C(t)\underset{\sim}{x}(t), \end{aligned} \tag{1}$$

where $A(t)$, $B(t)$ and $C(t)$ are $n\times n$ $(n = 202)$, $n\times p$ and $q\times n$ matrices of continuous or discontinuous functions of t, $\underset{\sim}{x}\varepsilon R^n$ is the state variable, $\underset{\sim}{u}\varepsilon R^p$ is the input variable, and $\underset{\sim}{y}\varepsilon R^q$ is the output variable. As such, the model will be suitable for study of small perturbations of mineral regimes in the ecosystem comprehended in terms of the 202 compartments.

Problem. The prototype ecosystem is unique, but its model obviously will not be. What consequences does this hold for the ecologists' objective of understanding mineral cycling in the actual ecosystem? Is the ecologist constrained to understand mainly the model? Must he construct a new model for each shade of gray in his set of objectives?

These problems become acute when data requirements are considered. The most pernicious cause of model failure in ecology has been inadequate data. Data accrue in ecological studies through costly and painstaking measurement.

Accurate and precise data are required for both identification and validation. The 202-compartment watershed model above could accomodate time series data for 23 inputs (Table I), 202 state variables, and approximately 8000 time-varying coefficients, clearly a herculean task for even a large team of scientists for a long period of time. The infeasibility of such an approach should be apparent.

Experience has shown that three classes of data are generally associated with an ecosystem model: (1) a small set of adequate time series, particularly for inputs, (2) a somewhat larger set of good point measurements (static), but not time series, and (3) an enormous set of parameters and variables that have never been measured, or for which only a few tentative estimates are available.

Problem. Are the data requirements of macromodels crippling, or are there ways around the problem? What does systems science have to say about minimal data for a model? For example, given a set of aggregated components and an interconnection topology, is there a critical data set where type (1) measurements above are essential? What is the pattern of model convergence to valid behavior as data of improved quality are added? Does this pattern admit the possibility of data supplementation by a model, i.e., the generation of a priori valid time series of parameters and variables?

Summarizing the two main points of this section, the ecologist--a pragmatist and not a theorist--will look to the systems scientist for guarantees that ecosystem models depict properties of actual ecosystems rather than model mathematics, and secondly, will be attentive to suggestions on how to reduce data requirements without sacrificing details of ecosystem processes.

III. Ecological Relevance of Systems Analysis Concepts

Total ecosystem modeling and systems analysis are new ventures in ecology. Most ecologists view them with skepticism and disdain. For example, the following views were published by the British Ecological Society and thought to have sufficient general appeal that they were later republished for U.S. consumption by the Ecological Society of America: "[One development] in ecological thought [has] debased our values in ecology..., the use, or, rather, the misuse of the computer model for whole systems analysis***

There is not even a problem to solve in whole systems analysis, merely complex numerical statement of what can be seen to be***Now I find the spectacle of a world's surface studded with groups of ecologists, engaged in massive quantification of the obvious, depressing in the extreme*** Whole systems analysis is irrelevant to our pressing problems***If some ecologists return to the fold at the end of it all, knowing what is in the black box, and if they then apply their knowledge to real problems perhaps it will not be too late." (Gifford, 1971, 1972).

Such widespread sentiments notwithstanding, the track of ecology now crosses that of systems science in the form of the ecosystem concept. As the current state of both fields encapsulates their past histories, there now must ensue a clash of paradigms to be resolved in the process of bisociative (Kestin, 1970) mixing of frames of reference. We in ecology who accept the tenets of systems thinking (Emery, 1969) recognize that it is our science that is immature, informal and non-rigorous, and that much of the methodology for elucidating ecosystems already exists in systems science. To avail ourselves of this body of theory and method it will be necessary to express ecosystems in terms that the systems scientist understands. Thus, the systems specialist has a set of automatic reflex responses when he sees Eqs. (1), whereas to the ecologist they are meaningless, and to suggest that they represent his complex ecosystem must somehow amuse him.

The best way to make ecology available to the systems scientist is through a systems model. A total ecosystem model of the scope of the watershed system outlined in Table I has never been completed, and most models that have been produced are exploratory, small scale, and both ecologically and mathematically trivial. One model of intermediate size is available that successfully simulates biomass dynamics of a shortgrass prairie in northeastern Colorado (Patten, 1973). A detailed description of this model is beyond the scope of this paper, but such a model could provide a point of focus for "closing the gap between theory and practice," one of the specific charges to the seminar.* It contains sufficient biological and environmental detail to be of genuine ecological interest, and it

*The model was made available to seminar participants at Sorrento.

belongs to a mathematical class of systems that makes it readily amenable to systems analysis.

Problem. What are the stability, observability, controllability and sensitivity characteristics of the abovementioned grassland model, and of what ecological significance are these properties to the prototype prairie ecosystem?

IV. Sensitivity Analysis for Ecosystems

It was suggested in the Introduction that, just as ecology will be changed by the systems approach, concepts and methods of systems analysis will also undergo adaptation as they are applied to macrosystems. This section presents some preliminary attempts to modify sensitivity analysis for application to ecosystems. Ecosystems are construed as a collection of storage "compartments", each described by a state variable, and an interactive topology. For purposes of exposition the linear model of Eqs. (1) will be assumed.

Let $\underset{\sim}{p} \varepsilon R^r$ be a vector of parameters that alters the system behavior. Then Eqs. (1) can be modified to

$$\dot{\underset{\sim}{x}}(t) = A(\underset{\sim}{p})\underset{\sim}{x}(t) + B(\underset{\sim}{p})\underset{\sim}{u}(t),$$

$$\underset{\sim}{y}(t) = C(\underset{\sim}{p})\underset{\sim}{x}(t). \qquad (2)$$

Define $\underset{\sim}{\sigma}_k(t)$ and $\underset{\sim}{\rho}_k(t)$ as state and output sensitivities, respectively, with respect to the *kth* parameter, p_k. Normally (e.g., Tomovic, 1963), these sensitivities are

$$\underset{\sim}{\sigma}_k(t) = \partial\underset{\sim}{x}/\partial p_k,$$

$$\underset{\sim}{\rho}_k(t) = \partial\underset{\sim}{y}/\partial p_k, \qquad (3)$$

leading to state and output sensitivity operators as follows:

$$\frac{\partial}{\partial p_k}[\dot{\underset{\sim}{x}}] = \frac{\partial}{\partial p_k}[A\underset{\sim}{x} + B\underset{\sim}{u}],$$

$$\dot{\underset{\sim}{\sigma}}_k = A\underset{\sim}{\sigma}_k + A'\underset{\sim}{x} + B\underset{\sim}{u}' + B'\underset{\sim}{u}; \qquad (4)$$

$$\frac{\partial}{\partial p_k}[\underset{\sim}{y}] = \frac{\partial}{\partial p_k}[C\underset{\sim}{x}],$$

$$\underset{\sim}{o}_k = C\underset{\sim}{\sigma}_k + C'\underset{\sim}{x}. \tag{5}$$

Primes indicate differentiation with respect to p_k. If input is independent of p_k then $\underset{\sim}{u}' = \underset{\sim}{0}$ and (4) is modified accordingly. The state sensitivity operator is obtained by solution of (4).

These sensitivity operators are used to approximate the model solution when a parameter is perturbed in a neighborhood around its nominal value (Pridgen, 1970). With $\underset{\sim}{x}(t)$ and $\underset{\sim}{y}(t)$ being the solution when p_k takes on its normal value p_k^{ν}, and $\underset{\sim}{x}(t,p_k^{\tau})$ and $\underset{\sim}{y}(t,p_k^{\tau})$ being the true solution when p_k has the perturbed value p_k^{τ}, then

$$\underset{\sim}{x}(t,p_k^{\tau}) \doteq \underset{\sim}{x}(t) + \underset{\sim}{\sigma}_k(t)\Delta p_k,$$
$$\underset{\sim}{y}(t,p_k^{\tau}) \doteq \underset{\sim}{y}(t) + \underset{\sim}{o}_k(t)\Delta p_k, \tag{6}$$

where $\Delta p_k = p_k^{\tau} - p_k^{\nu}$.

For ecosystem studies more than parametric sensitivity is required since the ecologist is not only interested in manipulating ecosystems but also in understanding their "sensitivity structure". Tomovic (1963), in fact, suggested that sensitivity theory should be extended to other system features than parameters. Three classes of sensitivities can be recognized as important for ecological studies:

1. State sensitivity,
2. Input sensitivity,
3. Parameter sensitivity.

This list is in approximate order of accessibility to ecologists for experimental and management manipulations.

A major difference between engineering systems and ecosystems is inability of the investigator to fix the latter in time while making a manipulation. Ecosystems, as they are being altered, always continue in motion, and the effects of perturbing a state, input or parameter must always be evaluated in the context of this motion. Thus, causal sensitivity is confounded with time-correlated sensitivity, making the concepts represented by Eqs. (3) inadequate for ecosystems.

Let the model of Eqs. (1) be rewritten

$$\dot{\underset{\sim}{x}}(t) = A(t)\underset{\sim}{x}(t) + \underset{\sim}{z}(t),$$
$$\underset{\sim}{y}(t) = \underset{\sim}{x}(t), \tag{7}$$

where $\underset{\sim}{z} = B\underset{\sim}{u}$. The ith compartment has the equation

$$\dot{x}_i(t) = \sum_{j=1}^{n} a_{ij}(t)x_j(t) + z_i(t),$$

or in functional form,

$$x_i = x_i[x_i(t_o), \underset{\sim}{x}(t), z_i(t), a_{ij}(t), t], \text{ all } j, \tag{8}$$

where $x_i(t_o)$ is the initial state, and z_i and the a_{ij}'s are independent of $\underset{\sim}{x}$ and of each other. The total differential is

$$dx_i = \frac{\partial x_i}{\partial x_i(t_o)} dx_i(t_o) + \sum_{\substack{j=1\\ j\neq 1}}^{n} \frac{\partial x_i}{\partial x_j} dx_j + \frac{\partial x_i}{\partial z_i} dz_i + \sum_{j=1}^{n} \frac{\partial x_i}{\partial a_{ij}} da_{ij} + \frac{\partial x_i}{\partial t} dt. \tag{9}$$

Letting p_k represent either a state variable x_k, input variable z_k, or parameter a_{ik}, a general expression for total derivatives is

$$\frac{dx_i}{dp_k} = \frac{\partial x_i}{\partial x_i(t_o)} \frac{dx_i(t_o)}{dp_k} + \sum_{\substack{j=1\\ j\neq 1}}^{n} \frac{\partial x_i}{\partial x_j}\frac{dx_j}{dp_k} + \frac{\partial x_i}{\partial z_i}\frac{dz_i}{dp_k} + \sum_{j=1}^{n} \frac{\partial x_i}{\partial a_{ij}}\frac{da_{ij}}{dp_k} + \frac{\partial x_i}{\partial t}\frac{dt}{dp_k}. \tag{10}$$

Embodied in this expression are three levels of sensitivities, which can best be clarified by means of an example.

A hypothetical system of four compartments is described by

$$\begin{bmatrix} \dot{x}_i \\ \dot{x}_j \\ \dot{x}_k \\ \dot{x}_\ell \end{bmatrix} = \begin{bmatrix} 0 & a_{ij} & a_{ik} & 0 \\ 0 & -a_{ij} & a_{jk} & a_{j\ell} \\ 0 & 0 & -(a_{ik}+a_{jk}) & 0 \\ 0 & 0 & 0 & -a_{j\ell} \end{bmatrix} \begin{bmatrix} x_i \\ x_j \\ x_k \\ x_\ell \end{bmatrix} + \begin{bmatrix} 0 \\ 0 \\ z_k \\ z_\ell \end{bmatrix}. \tag{11}$$

The total derivative of, say, x_i with respect to, e.g., z_k is formed as follows.

Each state variable is a function with the following arguments:

$$x_i = x_i[x_i(t_o), x_j(t), x_k(t), a_{ij}(t), a_{ik}(t), t], \tag{12}$$

$$x_j = x_j[x_j(t_o), x_k(t), x_\ell(t), a_{ij}(t), a_{jk}(t), a_{j\ell}(t), t], \tag{13}$$

$$x_k = x_k[x_k(t_o), z_k(t), a_{ik}(t), a_{jk}(t), t], \tag{14}$$

$$x_\ell = x_\ell[x_\ell(t_o), z_\ell(t), a_{j\ell}(t), t]. \tag{15}$$

Hence,

$$\frac{dx_i}{dz_k} = \frac{\partial x_i}{\partial x_i(t_o)} \frac{dx_i(t_o)}{dz_k} + \frac{\partial x_i}{\partial x_j}\frac{dx_j}{dz_k} + \frac{\partial x_i}{\partial x_k}\frac{dx_k}{dz_k} + \frac{\partial x_i}{\partial a_{ij}}\frac{da_{ij}}{dz_k} + \frac{\partial x_i}{\partial a_{ik}}\frac{da_{ik}}{dz_k} + \frac{\partial x_i}{\partial t}\frac{dt}{dz_k}, \tag{16}$$

or, substituting similar expressions for dx_j/dz_k and dx_k/dz_k, and dropping the initial state terms whose values are zero,

$$\begin{aligned}
\frac{dx_i}{dz_k} = {} & \left(\frac{\partial x_i}{\partial x_k}\frac{\partial x_k}{\partial z_k}\right) + \left(\frac{\partial x_i}{\partial x_j}\frac{\partial x_j}{\partial x_k}\frac{\partial x_k}{\partial z_k}\right) + \left(\frac{\partial x_i}{\partial x_j}\frac{\partial x_j}{\partial x_\ell}\frac{\partial x_\ell}{\partial z_\ell}\right)\frac{dz_\ell}{dz_k} \\
& + \left(\frac{\partial x_i}{\partial a_{ik}} + \frac{\partial x_i}{\partial x_k}\frac{\partial x_k}{\partial a_{ik}} + \frac{\partial x_i}{\partial x_j}\frac{\partial x_j}{\partial a_{ik}} + \frac{\partial x_i}{\partial x_j}\frac{\partial x_j}{\partial x_k}\frac{\partial x_k}{\partial a_{ik}}\right)\frac{da_{ik}}{dz_k} \\
& + \left(\frac{\partial x_i}{\partial a_{ij}} + \frac{\partial x_i}{\partial x_j}\frac{\partial x_j}{\partial a_{ij}}\right)\frac{da_{ij}}{dz_k} + \left(\frac{\partial x_i}{\partial x_k}\frac{\partial x_k}{\partial a_{jk}} + \frac{\partial x_i}{\partial x_k}\frac{\partial x_j}{\partial x_k}\frac{\partial x_k}{\partial a_{jk}}\right)\frac{da_{jk}}{dz_k} \\
& + \left(\frac{\partial x_i}{\partial x_j}\frac{\partial x_j}{\partial a_{j\ell}} + \frac{\partial x_i}{\partial x_j}\frac{\partial x_j}{\partial x_\ell}\frac{\partial x_\ell}{\partial a_{j\ell}}\right)\frac{da_{j\ell}}{dz_k} \\
& + \left(\frac{\partial x_i}{\partial t} + \frac{\partial x_i}{\partial x_k}\frac{\partial x_k}{\partial t} + \frac{\partial x_i}{\partial x_j}\frac{\partial x_j}{\partial t} + \frac{\partial x_i}{\partial x_j}\frac{\partial x_j}{\partial x_k}\frac{\partial x_k}{\partial t} + \frac{\partial x_i}{\partial x_j}\frac{\partial x_j}{\partial x_\ell}\frac{\partial x_\ell}{\partial t}\right)\frac{dt}{dz_k}.
\end{aligned} \tag{17}$$

The three levels of sensitivity referred to above are:

1. *Direct Causal Sensitivity*, $\partial \underset{\sim}{x}/\partial p_k$

This represents the sensitivity of state variables to other variables, inputs or coefficients that are directly coupled to the dependent variable in the system topology. Indirect causal effects and correlated effects of system motion are not included. This is the sensitivity that was previously defined in Eq. (3). In Eq. (17) $\partial x_i/\partial z_k$ is zero and does not appear since there is no direct coupling of x_i to z_k.

2. *Total (Direct and Indirect) Causal Sensitivity,* $\delta \underset{\sim}{x}/\delta p_k$

This represents state sensitivty mediated over all paths, both direct and indirect, in the system interconnection leading from the perturbed parameter to the dependent variable. Correlated effects of dynamic motion are not included. The first two terms of Eq. (17) comprise $\delta x_i/\delta z_k$. For applications to ecosystems, which have complex patterns of causal interactions, both with and without circuits, this is the sensitivity measure of choice. The direct causal measure above and of Eq. (3) is only one component of total causal sensitivity.

3. *Total (Causal and Correlated) Sensitivity,* $d\underset{\sim}{x}/dp_k$

Causal sensitivity, an attribute of systems models, is not generally measureable in the real ecosystem. It is confounded in the system's dynamic motion, which continues whether or not a perturbation experiment is performed. The remaining terms of Eq. (17) after the second represent correlative contributions to total sensitivity. These influences cannot be stopped in order to measure $\delta x_i/\delta p_k$. Hence, total sensitivity, including both causal and correlated effects, is the only practical sensitivity measure presently available for ecosystems. Despite its shortcomings as a measure of sensitivity it has the advantage of being model free. That is, being a total derivative, total sensitivity can be computed as a ratio of two time derivatives:

$$\frac{d\underset{\sim}{x}}{dp_k} = \frac{\dot{\underset{\sim}{x}}}{\dot{p}_k} \quad . \qquad (18)$$

No simulation model is needed to generate this information.

The total sensitivity acts as an operator to map rates of change of states, inputs or parameters into rates of change of dependent state variables:

$$\dot{x}_i = \left(\frac{dx_i}{dp_k}\right)\dot{p}_k. \tag{19}$$

For example, state transitions in nominal motion are given by

$$x_i(t+dt) = x_i(t) + (\dot{x}_i)dt, \tag{20}$$

or, substituting Eq. (19),

$$x_i(t+dt) = x_i(t) + \left(\frac{dx_i}{dp_k}\right)\dot{p}_k dt. \tag{21}$$

This may be rewritten

$$x_i(t+dt) = x_i(t) + \left(\frac{dx_i}{dp_k}\right)dp_k. \tag{22}$$

For larger perturbations, Δp_k, a first order approximation is

$$x_i(t+dt) \doteq x_i(t) + \left(\frac{dx_i}{dp_k}\right)\Delta p_k. \tag{23}$$

Thus, for ecosystems (and other richly connected macro-systems), the appropriate sensitivity operators are not those defined by Eqs. (3), but instead

$$\begin{aligned} \underset{\sim}{\sigma}_k(t) &= d\underset{\sim}{x}/dp_k, \\ \underset{\sim}{\rho}_k(t) &= d\underset{\sim}{y}/dp_k, \end{aligned} \tag{24}$$

and Eq. (6) is understood as modified accordingly.

V. Control in Ecosystems

An ecologically interesting result arises immediately from the use of total sensitivity in ecosystem analysis. Let v(t) and w(t) be any pair of system state variables, inputs or parameters that are *mutually causally related*,

$$\frac{\delta v}{\delta w} \neq 0, \ \frac{\delta w}{\delta v} \neq 0,$$

in a system interconnection.* Then, their total sensitivities must be inverses:

$$\frac{dv}{dw} = \frac{1}{dw/dv} . \tag{25}$$

If the instantaneous effect of v(t) on w(t) is small (or large), then the effect of w(t) on v(t) must be large (small). This amounts to a *reciprocity principle of control*.

Application of this obvious truism to ecosystems implies that plant and animal species, or other ecosystem components, that achieve relative independence (low sensitivity) from other components automatically, so long as causal connection is maintained, become capable of exerting a large ultimate influence on those components. Conversely, parameters or variables that have small effects on others are subject to control by the latter. This principle holds also, obviously, for any general system. In the human sphere, probably every successful businessman, politician and other man of influence who has ever lived has somehow grasped it and managed to put it to use.

If the idea is understood in other fields, it it new in this general form to ecology. Ecologists have long debated the significance in ecosystems of bulk processors f energy and materials, such as primary producers and microorganisms, compared to carnivores and other animals at the tops of food chains that seem to have a regulatory function that far exceeds the energy and matter they handle. Parasites, pathogens and man are all good examples of organisms that exist on the energetic and material fringes of ecosystems, relatively immune to ecological reshufflings of even major proportions, but capable nevertheless of exerting profound influence. Apparently, as with the brain and man-made controllers, these organisms serve a large information function --system regulation-- while exerting small demands for energy and material. The reciprocity principle embodied in Eq. (25) endows them with their power and at the same time

*In the ecosystem, state variables are all mutually causal due to nesting within grand loops provided by the material cycling phenomenon. This is an important corollary of recycling from a systems analysis point of view.

protects them in their vital regulatory functions from undue stresses generated both within and outside the ecosystem. The implications for species evolution in the context of evolving ecosystems are profound; controlling species must disconnect from the ecosystem to the point of reduced sensitivity, but not to the point of severing their causal influence.

A mind/body or controller/plant separation based on the principle may be intrinsic to all organizations. In respect to ecosystems, several variable structure aspects seem especially germane here:

1. The total sensitivity structure of an ecosystem is itself continuously in motion with the motion of the system. Relations between organisms and environmental components are thus not fixed, and any experimental or management scheme based on such a view is bound to be erroneous. Variable, rather than fixed, control relations within the ecosystem make a rigid controller/controlled separation of system components, as in classical control theory, totally inappropriate. Control relationships are dynamic, and any complete theory of ecosystems will have to allow for the variable and potentially even ephemeral nature of both control and process components. What may at one time be controlled may at another be controlling.

2. Remote control may be the most powerful kind. Adjacent components in an ecosystem interconnection rarely seem to have enough amplification (sensitivity) to influence one another significantly. It is only in cascades of coupled variables that significant amplifying power is developed, in accordance with the multiplication principle of serial interactions. In the context of classical ecological doctrines about prey regulation by their immediate predators, remote control is one of the most compelling types of generalizations that makes the systems approach mandatory for studying ecosystems.

Acknowledgements. Jack B. Waide, Jackson R. Webster and Jeffrey J. Lee aided in identifying and clarifying the three levels of sensitivity recognized in this paper.

REFERENCES

Emery, F. E. (ed.). 1969. Systems Thinking, Baltimore, Maryland. Penguin.

Gifford, D. 1971. Bull. British Ecol. Soc. 2: 2.

______ 1972. Bull. Ecol. Soc. Amer. 53: 9.

Kestin, J. 1970. Amer. Scient. 58: 250-256.

Patten, B. C. 1970. Simulation 15: 264-268.

______ (ed.). 1973. Systems Analysis and Simulation in Ecology, Vol. III. A State Space Model for Grassland. New York. Academic Press. In preparation.

Tomovic, R. 1963. Sensitivity Analysis of Dynamic Systems. New York. McGraw-Hill.

Ziegler, B. P. 1970. Towards a formal theory of modeling and simulation. U. Mich. Tech. Rep. 032960-5-T.

______ 1971a. Towards a formal theory of modeling and simulation. U. Mich. Tech. Rep. 032960-15-T.

______ 1971b. Modeling and simulation: Structure preserving relations for continuous and discrete time systems. U. Mich. Tech. Rep. 032960-11-T.

______ and Weinberg, R. 1970. J. Theor. Biol. 29: 35-56.

TABLE I

Coweeta Watershed Model Compartments

Designation		Description	
Plant Submodel			
X1	Active*	Immature	White Oaks
X2	Inactive**	"	"
X3	Leaves	"	"
X4	Active	Mature	"
X5	Inactive	"	"
X6	Leaves	"	"
X7	Reproductive***	"	"

* Active = active woody tissue.
** Inactive = inactive woody tissue.
*** Reproductive = flowers & fruits.

X8	Active	Immature	Black Oaks
X9	Inactive	"	"
X10	Leaves	"	"
X11	Active	Mature	"
X12	Inactive	"	"
X13	Leaves	"	"
X14	Reproductive	"	"
X15	Active	Immature	Hickories
X16	Inactive	"	"
X17	Leaves	"	"
X18	Active	Mature	"
X19	Inactive	"	"
X20	Leaves	"	"
X21	Reproductive	"	"
X22	Active	Immature	Tulip Poplar
X23	Inactive	"	"
X24	Leaves	"	"
X25	Active	Mature	"
X26	Inactive	"	"
X27	Leaves	"	"
X28	Reproductive	"	"
X29	Active	Immature	Red Maple
X30	Inactive	"	"
X31	Leaves	"	"
X32	Active	Mature	"
X33	Inactive	"	"
X34	Leaves	"	"
X35	Reproductive	"	"
X36	Active	Immature	Dogwood
X37	Inactive	"	"
X38	Leaves	"	"
X39	Active	Mature	"
X40	Inactive	"	"
X41	Leaves	"	"
X42	Reproductive	"	"
X43	Active	Immature	Subdominant species
X44	Inactive	"	"
X45	Leaves	"	"
X46	Active	Mature	"
X47	Inactive	"	"
X48	Leaves	"	"
X49	Reproductive	"	"
X50	Active	Immature	Conifers
X51	Inactive	"	"

X52	New leaves	"	"
X53	Old leaves	"	"
X54	Active	Mature	"
X55	Inactive	"	"
X56	New leaves	"	"
X57	Old leaves	"	"
X58	Reproductive	"	"
X59	Active		*Kalmia, Rhododendron*
X60	Inactive		"
X61	New leaves		"
X62	Old leaves		"
X63	Reproductive		"
X64	Active		Deciduous Shrubs
X65	Inactive		"
X66	Leaves		"
X67	Reproductive		"
X68	Stems and leaves		Perennial Herbs
X69	Reproductive		"
X70	Stems and leaves		Annual Herbs
X71	Reproductive		"
X72	All parts		Epiphytic mosses, lichens, algae, etc.
X73	All parts		Forest floor mosses, lichens, algae, etc.

Vertebrate Submodel

X74	Small migratory birds
X75	Hawks and owls
X76	Scavenger birds
X77	Gallinaceous birds
X78	Mice
X79	Shrews
X80	Squirrels
X81	Moles
X82	Predaceous and omnivorous mammals
X83	Small herbivorous

	mammals
X84	Bats
X85	Large herbivorous mammals
X86	Snakes and lizards
X87	Turtles
X88	Salamanders
X89	Frogs and toads
Canopy Invertebrate Submodel	
X90	Orthoptera
X91	Aphids
X92	Leafhoppers and Hemiptra
X93	Lepidoptera larvae
X94	Lepidoptera adults
X95	Ants
X96	Inflorescence feeders
X97	Leaf-eating Cole-optera
X98	Terminal tissue feeders
X99	Snails
X100	Leaf miners
X101	Bark and wood borers
X102	Seed and ovule feeders
X103	Canopy epiphyte grazers
X104	Forest floor li-chen-moss grazers
X105	Parasites
X106	Web spiders
X107	Non-web spiders
X108	Coccinellidae, etc.
X109	Neuroptera
X110	Hemiptera
X111	Diptera and

	Mecoptera
X112	Hymenoptera
X113	Odonata
X114	Phalangids
X115	Mantids
X116	Honeydew
Stream Submodel	
X117	Coarse detritus
X118	Coarse detritus fungi
X119	Coarse detritus bacteria
X120	Fine detritus
X121	Fine detritus fungi
X122	Fine detritus bacteria
X123	Solution
X124	Solution bacteria
X125	Particulate in-organic matter
X126	Phytoplankton
X127	Microfauna
X128	Periphyton
X129	Carrion
X130	Carrion fungi
X131	Carrion bacteria
X132	Uncommon insects
X133	Crayfish
X134	*Peltoperla*
X135	Cranefly
X136	Filter feeders
X137	Grazers
X138	Root eaters
X139	*Lanthus, Rhyacophila*
X140	Salamanders, frogs, fish
X141	Non-feeding stream emergent insects

Decomposer Submodel	
X142	Coarse leaf litter
X143	Coarse-litter fungi
X144	Coarse-litter bacteria
X145	Fine leaf litter
X146	Fine-litter fungi
X147	Fine-litter bacteria
X148	Snails
X149	Diplopods and isopods
X150	Mites (excluding predatory forms)
X151	Collembola
X152	Nematodes (free-living)
X153	Protozoa
X154	Orthoptera
X155	Immature saprohagous and mycophagous insects
X156	Micropredators
X157	Macropredators
X158	Earthworms
X159	Ants
X160	Leaf humus
X161	Humus fungi
X162	Humus bacteria
X163	Non-feeding pupae in litter
X164	Log-branch phloem
X165	Log-branch heartwood
X166	Log-branch fungi
X167	Log-branch bacteria
X168	Log-branch Coleoptera
X169	Isoptera
X170	Log-branch Blattidae

X171	Passalidae
X172	Other log-branch borers
X173	Diptera
X174	Passalid bettle feces
X175	Passalid feces fungi
X176	Passalid feces bacteria
X177	Feces
X178	Feces bacteria
X179	Feces fungi
X180	Feces Blattidae
X181	Feces Coleoptera
X182	Carrion
X183	Carrion fungi
X184	Carrion bacteria
X185	Carrion Coleoptera
X186	Dermaptera
X187	Live roots
X188	Mycorrhizal fungi
X189	Rhizospheral bacteria
X190	Dead roots
X191	Dead-root fungi
X192	Dead-root bacteria
X193	Scarab beetles
X194	Cicadas (and other live root suckers)
X195	Parasitic nematodes
X196	Litter solution
X197	Topsoil
X198	Topsoil solution
X199	Subsoil
X200	Subsoil solution
X201	Immobilized ions and soil minerals
X202	Groundwater

External Inputs to System (Z)

Z1	Wet precipitation
Z2	Dry precipitation
Z3	Immigration
Z4	External food sources
Z5	Rock weathering

External Outputs from System (Y)

Y1	Emigration
Y2	External defecation
Y3	Particulate organic
Y4	Dissolved organic and inorganic matter
Y5	Particulate inorganic
Y6	Harvest

External Control Factors

Time (t)	pH(PH)
Air temperature (TA)	Inhibitory substances in live vegetation (IL)
Soil temperature (TS)	Inhibitory substances on litter soil (IS)
Water temperature (TW)	Photoperiod (PP)
Rainfall (PT)	Egg production (EP)
Humidity (HU)	Hibernation - aestivation (HI)
Soil moisture (MS)	Parasites - pathogens (PA)
Litter moisture (ML)	Stream flow rate (FR)
Wind (WI)	Carbon : nitrogen ratio (CN)
Nutrient concentration in water (DN)	